SPACE SCIENCE, EXPLORATION AND POLICIES

MAJOR PROJECTS AT NASA

SELECT ASSESSMENTS

SPACE SCIENCE, EXPLORATION AND POLICIES

Additional books in this series can be found on Nova's website
under the Series tab.

Additional E-books in this series can be found on Nova's website
under the E-book tab.

GOVERNMENT PROCEDURES AND OPERATIONS

Additional books in this series can be found on Nova's website
under the Series tab.

Additional E-books in this series can be found on Nova's website
under the E-book tab.

SPACE SCIENCE, EXPLORATION AND POLICIES

MAJOR PROJECTS AT NASA

SELECT ASSESSMENTS

ELOY D'ARAMITZ
EDITOR

nova publishers
New York

Library of Congress Cataloging-in-Publication Data

ISBN: 978-1-62417-866-5

Published by Nova Science Publishers, Inc. † New York

CONTENTS

PREFACE

This book provides a snapshot of how well NASA is planning and executing its major acquisitions--an area that been on the Government Accountability Office's (GAO's) high risk list since 1990. Over the past 4 years, a range of projects such as robotic probes designed to explore the Martian surface and telescopes intended to explore the universe have been reviewed. NASA has launched many of the projects and most of the projects are returning science results as intended. For example, the Dawn spacecraft reached the solar system's second largest asteroid, Vesta, in July 2011 and is beaming back images and data to scientists about how the solar system formed. Another project called GRAIL, launched in September 2011, and its twin spacecraft recently entered orbit around the moon to begin its measurements of lunar gravity.

Chapter 1 – This is GAO’s fourth annual assessment of the National Aeronautics and Space Administration’s (NASA) large-scale projects. GAO’s work has shown that these projects, while producing ground-breaking research and advancing our understanding of the universe, tend to cost more and take longer to develop than planned, and are often approved without evidence of a sound business case. GAO has designated NASA’s acquisition management a high risk area.

In response to congressional direction, GAO reviewed NASA's major projects. Specifically, this report provides observations about the performance of NASA's major projects, assesses knowledge attained at key junctures of development, identifies challenges that can contribute to cost and schedule growth, and outlines steps NASA is taking to improve its acquisitions. To conduct this review, GAO assessed data on 21 projects with an estimated life-cycle cost of over $250 million, including data on projects’ cost, schedule, technology maturity, design stability, and contracts; analyzed monthly project status reports; and interviewed NASA and contractor officials. GAO also reviewed project cost estimates and interviewed officials responsible for NASA’s cost estimation policy.

Chapter 2 – JWST is one of NASA’s most expensive and technologically advanced science projects, intended to advance understanding of the origin of the universe. In 2011, JWST was rebaselined with a life cycle cost estimate of $8.8 billion and a launch readiness date in October 2018 — almost nine times the cost and more than a decade later than originally projected in 1999. Concern about the magnitude of JWST's cost increase and schedule delay and their effects on NASA’s progress on other high-priority missions led conferees for the Consolidated and Further Continuing Appropriations Act, 2012, to direct GAO to report on the project. Specifically, GAO assessed (1) the extent to which NASA’s revised cost and schedule estimates are reliable based on best practices, (2) the major risks and technological challenges JWST faces, and (3) the extent to which NASA has improved oversight of JWST. To do this, GAO compared NASA's revised cost and schedule estimates with best practice criteria, reviewed relevant contractor and NASA documents, and interviewed project and contractor officials.

In: Major Projects at NASA: Select Assessments
Editor: Eloy D'Aramitz
ISBN: 978-1-62417-866-5

Chapter 1

NASA: Assessments of Selected Large-Scale Projects*

United States Government Accountability Office

Why GAO Did This Study

This is GAO's fourth annual assessment of the National Aeronautics and Space Administration's (NASA) large-scale projects. GAO's work has shown that these projects, while producing ground-breaking research and advancing our understanding of the universe, tend to cost more and take longer to develop than planned, and are often approved without evidence of a sound business case. GAO has designated NASA's acquisition management a high risk area.

In response to congressional direction, GAO reviewed NASA's major projects. Specifically, this report provides observations about the performance of NASA's major projects, assesses knowledge attained at key junctures of development, identifies challenges that can contribute to cost and schedule growth, and outlines steps NASA is taking to improve its acquisitions. To conduct this review, GAO assessed data on 21 projects with an estimated life-cycle cost of over $250 million, including data on projects' cost, schedule, technology maturity, design stability, and contracts; analyzed monthly project status reports; and interviewed NASA and contractor officials. GAO also reviewed project cost estimates and interviewed officials responsible for NASA's cost estimation policy.

What GAO Recommends

GAO is not making any new recommendations in this report, but has made prior recommendations to address transparency in project costs and the lack of consistent design metrics; NASA concurred and is taking steps to address them.

---

* This is an edited, reformatted and augmented version of the Highlights of GAO-12-207SP, a report to congressional committees, dated March 2012.

What GAO Found

GAO assessed 21 NASA projects with a combined life-cycle cost that exceeds $43 billion. Of those 21 projects, 6 were in an early phase of development called formulation, and 15 had entered the implementation phase where cost and schedule baselines were established. Five of the 15 projects in implementation successfully launched in 2011, and two of them met their cost and schedule baselines. NASA's largest science project—the James Webb Space Telescope (JWST)—however, has experienced development cost growth of $3.6 billion—or 140 percent—and a schedule delay of over 4 years. While the development cost and schedule for most of the projects in implementation remained relatively stable, the impact of the JWST increases on the portfolio is significant. For example, 14 of the 15 projects currently in implementation, excluding JWST, had an average development cost growth of $79 million—or 14.6 percent—and schedule growth of 8 months from their baselines. With JWST, these numbers increase dramatically to almost 47 percent and 11 months, respectively. Cost and schedule increases within NASA's most technologically advanced and costly projects, such as JWST, can have cascading effects on the rest of NASA's portfolio. For example, the administration has proposed to terminate funding for the joint NASA/ESA EMTGO project, and another large project in our review—the Mars Science Laboratory (MSL)—experienced substantial cost overruns that led NASA to take funding from other projects. MSL and JWST account for approximately $11.4 billion—or 51 percent—of the total life-cycle costs for the 15 projects in implementation during our review.

Most of the projects that GAO reviewed did not meet technology maturity and design stability best practices criteria, which if followed can lessen cost and schedule risks faced by the project. Specifically, 10 of the 16 projects that held a preliminary design review moved forward without first maturing technologies. In addition, 13 of the 14 projects that held a critical design review did so without first achieving design stability. Some projects reported using other methods to assess design stability. Many of the projects GAO reviewed for this report also experienced challenges in the areas of launch vehicles, contractor management, parts, development partner performance, and funding. For example, nine projects we reviewed reported challenges with launch vehicles, including their increasing cost and availability. New launch vehicles are in development, but have not yet been certified, and another vehicle has failed on its two most recent flights.

The agency is continuing its implementation of initiatives to reduce acquisition management risk. One prominent effort is the Joint Cost and Schedule Confidence Level (JCL), a new cost estimation tool that involves a probabilistic analysis of cost, schedule, and risk inputs to arrive at development cost and schedule estimates associated with various confidence levels.

Five projects GAO reviewed have completed a JCL. NASA officials stated a few projects have excluded or not fully considered relevant cost inputs and risks, such as launch vehicle costs. GAO was unable to confirm that the five projects that prepared estimates using the JCL were budgeted at the approved confidence level. NASA has not yet launched a project that prepared a JCL; therefore, NASA officials stated it will take several years to evaluate the impact and effectiveness of the JCL in improving cost and schedule estimating for its major projects.

ABBREVIATIONS

AFB	Air Force Base
AFS	Air Force Station
ASI	Argenzia Spaciale Italiana (Italian Space Agency)
ATLAS	Advanced Topographic Laser Altimeter System
CDDS	Cavity Door Drive System
CDR	critical design review
CONAE	Comision Nacional de Actividades Espaciales (Space Agency of Argentina)
CSA	Canadian Space Agency
DCI	data collection instrument
DPR	dual-frequency precipitation radar
DSN	Deep Space Network
DWSS	Defense Weather Satellite System
EDL	entry, descent, and landing
EEE	electrical, electronic, and electromechanical
EMTGO	ExoMars Trace Gas Orbiter
ESA	European Space Agency
EVM	earned value management
FAA	Federal Aviation Administration
FORCAST	Faint Object Infrared Camera for the SOFIA Telescope
FPI	Fast Plasma Investigation
GLAST	Gamma-ray Large Area Space Telescope
GMI	GPM microwave imager
GPM	Global Precipitation Measurement (mission)
GPS	Global Positioning System
GRAIL	Gravity Recovery and Interior Laboratory
GREAT	German Receiver for Astronomy at Terahertz Frequencies
HEPS	High Efficiency Power Supply
HOPE	Helium-Oxygen-Proton-Electron
ICESat-2	Ice, Cloud, and Land Elevation Satellite-2
IPO	Integrated Program Office
ISIM	Integrated Science Instrument Module
ISS	International Space Station
JADE I	Jovian Auroral Distributions Experiment–Ion I
JAXA	Japan Aerospace Exploration Agency
JCL	Joint Cost and Schedule Confidence Level
JIRAM	Jupiter InfraRed Auroral Mapper
JPL	Jet Propulsion Laboratory
JPSS	Joint Polar Satellite System
JWST	James Webb Space Telescope
KDP	key decision point
LADEE	Lunar Atmosphere and Dust Environment Explorer
LDCM	Landsat Data Continuity Mission
LLCD	Lunar Laser Com Demo

LRO	Lunar Reconnaissance Orbiter
MagEIS	Magnetic Electron Ion Spectrometer
MAVEN	Mars Atmosphere and Volatile EvolutioN
MDR	mission design review
MEP	Mars Exploration Program
MMS	Magnetospheric Multiscale
MPCV	Multi-Purpose Crew Vehicle
MSL	Mars Science Laboratory
NASA	National Aeronautics and Space Administration
NGIMS	Neutral Gas and Ion Mass Spectrometer
NID	NASA Interim Directive
NLS	NASA Launch Services
NOAA	National Oceanic and Atmospheric Administration
NPR	NASA Procedural Requirements
NPOESS	National Polar-orbiting Operational Environmental Satellite System
NPP	NPOESS Preparatory Project
NTIA	National Telecommunications and Information Administration
OCFO	Office of the Chief Financial Officer (NASA)
OCO	Orbiting Carbon Observatory
PDR	preliminary design review
PFR	Problem/Failure Report
PSE	Power System Electronics
RBSP	Radiation Belt Storm Probes
SAC	Satélite de Aplicaciones Científicas
SDO	Solar Dynamics Observatory
SID	Strategic Investments Division (NASA)
SIR	System Integration Review
SLS	Space Launch System
SMAP	Soil Moisture Active and Passive
SPP	Solar Probe Plus
SOFIA	Stratospheric Observatory for Infrared Astronomy
SRR	systems requirements review
TDRS	Tracking and Data Relay Satellite
TIRS	Thermal Infrared Sensor
TPS	Thermal Protection System
TRL	technology readiness level
USGS	U.S. Geological Survey
WISE	Wide-field Infrared Survey Explorer

March 1, 2012

Congressional Committees

This is GAO's fourth annual assessment of the National Aeronautics and Space Administration's (NASA) major projects. This report provides a snapshot of how well NASA is planning and executing its major acquisitions—an area that has been on GAO's high risk list since 1990. Over the past 4 years, our review has covered a range of projects, such as robotic probes designed to explore the Martian surface and telescopes intended to explore the universe. NASA has launched many of the projects included in our reviews, and most of these projects are returning science results as intended (See appendix V for more details). For example, the Dawn spacecraft reached the solar system's second largest asteroid, Vesta, in July 2011 and is beaming back images and data to scientists about how the solar system formed. Since our last report, NASA successfully launched five projects. One of those projects, GRAIL, launched in September 2011, and its twin spacecraft recently entered orbit around the moon to begin its measurements of lunar gravity. The past year, however, has also continued to be a turbulent one for NASA's human spaceflight program and its largest science project. The Constellation program, which was NASA's largest program and was preliminarily estimated to cost in the tens of billions of dollars, was cancelled after facing significant technical and funding issues. NASA spent most of the past year transitioning from the Constellation program to the new Orion Multi-Purpose Crew Vehicle and Space Launch System programs, as directed by Congress in the NASA Authorization Act of 2010.[1] Despite being required to report to Congress within 90 days of enactment of the NASA Authorization Act of 2010 on the designs for the Multi-Purpose Crew Vehicle and Space Launch System, the agency delayed for several months in order to perform cost estimates and study alternatives before issuing the report in September 2011. The cancellation of the Constellation program and transition to a new launch approach has likely increased the amount of time the United States will be without a human spaceflight capability. In addition, the James Webb Space Telescope project—NASA's largest science project—focused last year on responding to an independent review that reported problems with overall management, a lack of effective oversight, and a funding baseline that did not reflect the most probable cost and resulted in a project that was not executable. In addition to multiple changes to address management of the program, in October 2011, NASA announced a $3.7 billion cost increase and a launch delay of over 4 years that will have reverberating effects on the portfolio for years to come. Such issues continue to affect NASA's ability to conduct its ground-breaking work in an efficient and effective manner.

NASA has taken steps in recent years to help improve its acquisition management through several initiatives aimed at improving cost estimating and management oversight. While the overall outcomes of these efforts will take time to become apparent, NASA has indicated that it continues to be committed to the initiatives with the goal of improving performance. The explanatory statement of the House Committee on Appropriations accompanying the Omnibus Appropriations Act of 2009 directed GAO to prepare project status reports on selected large-scale NASA programs, projects, or activities.[2] This report responds to that mandate. Specifically, we assess (1) performance of NASA's major projects and the agency's management of those projects during development, (2) knowledge attained by key junctures in the acquisition process, (3) other challenges that can affect project execution, and (4) NASA's continued efforts to improve its acquisition management. The report expands on the importance of providing decision-makers with an independent, knowledge-based assessment of individual systems that identifies potential risks and allows them to take actions to put projects that are early in the development cycle in a better position to succeed. Our approach included an examination of the current phase of a project's development and of how the 21 projects were advancing. Each project we reviewed was in either the formulation phase or the implementation phase of the project life cycle, and had an estimated life-cycle cost of over $250 million.

In the formulation phase, the project defines requirements—what the project is being designed to do—matures technology, establishes a schedule, estimates costs, and produces a plan for implementation. In the implementation phase, the project carries out these plans, performing final design and fabrication as well as testing components and system assembly, integrating these components and testing how they work together, and launching the project. This phase also includes the period from a project's launch through mission completion. NASA provided updated cost and schedule data as of January 2012 for projects in implementation during our review, or 15 of the 21 major projects; the remaining 6 were in formulation. We reviewed and compared that data to previously established cost and schedule baselines. We assessed the 15 projects' cost and schedule and characterized growth as significant if it exceeded the thresholds that trigger cost or schedule growth reporting to the Congress under the law.[3] In addition, NASA provided cost and schedule information for projects that have launched and that have been reported on in our prior work—see appendix III for a listing of projects covered in previous reports. We assessed technology maturity and design stability using established criteria for knowledge-based acquisitions and other GAO work on system acquisitions.[4]

Additionally, as a result of our analysis of interviews with project officials and information provided by the projects, we identified other challenges—funding, launch vehicles, contractor management, parts, and development partner performance—that can affect project outcomes.

This list of challenges is not exhaustive, and we believe these challenges will evolve, as they have in previous years, as we continue this work in the future. We took appropriate steps to address data reliability, such as clarifying data discrepancies and corroborating NASA-generated data with other sources where applicable. The individual project offices were given an opportunity to provide comments and technical clarifications on our assessments prior to their inclusion in the final product, which were incorporated as appropriate.

Appendix II contains detailed information on our scope and methodology. We conducted this performance audit from March 2011 to March 2012 in accordance with generally accepted government auditing standards.

Those standards require that we plan and perform the audit to obtain sufficient, appropriate evidence to provide a reasonable basis for our findings and conclusions based on our audit objectives. We believe that the evidence obtained provides a reasonable basis for our findings and conclusions based on our audit objectives. We are not making recommendations in this report.

BACKGROUND

A Sound Business Case Underpins Successful Acquisition Outcomes

The development and execution of a knowledge-based business case for NASA's projects can provide early recognition of challenges, allow managers to take corrective action, and place needed and justifiable projects in a better position to succeed. Our studies of best practice organizations show the risks inherent in NASA's work can be mitigated by developing a solid, executable business case before committing resources to a new product's development.[5] In its simplest form, a knowledge-based business case is evidence that (1) the customer's needs are valid and can best be met with the chosen concept and that (2) the chosen concept can be developed and produced within existing resources—that is, proven technologies, design knowledge, adequate funding, adequate time, and adequate workforce to deliver the product when needed. A program should not be approved to go forward into product development unless a sound business case can be made. If the business case measures up, the organization commits to the development of the product, including making the financial investment. Our work examining best practices has shown that developing business cases based on matching requirements to resources before a program starts leads to more predictable program outcomes—that is, programs are more likely to be successfully completed within cost and schedule estimates and deliver anticipated system performance.[6]

At the heart of a business case is a knowledge-based approach to product development, a best practice among leading commercial firms. Those firms have created an environment and adopted practices that put their program managers in a good position to meet expectations. A knowledge-based approach requires that managers demonstrate high levels of knowledge as the program proceeds from technology development to system development and, finally, production. In essence, knowledge reduces risk over time. This building of knowledge can be described over the course of a program as follows:

- When a project begins development, the customer's needs should match the developer's available resources—mature technologies, time, and funding. An indication of this match is the demonstrated maturity of the technologies required to meet customer needs—referred to as critical technologies. If the project is relying on heritage—or pre-existing—technology, that technology must be in the appropriate form, fit, and function to address the customer's needs within available resources. The project will generally enter development after completing the preliminary design review, at which time a business case should be in hand.
- Then, about midway through the project's development, its design should be stable and demonstrate it is capable of meeting performance requirements. The critical design review takes place at that point in time because it generally signifies when the program is ready to start building production-representative prototypes. If project development continues without design stability, costly re-designs to address changes to project requirements and unforeseen challenges can occur.
- Finally, by the time of the production decision, the product must be shown to be producible within cost, schedule, and quality targets and have demonstrated its reliability, and the design must demonstrate that it performs as needed through realistic system-level testing. Lack of testing increases the possibility that project managers will not have information that could help avoid costly system failures in late stages of development or during system operations.

Our work examining best practices has identified numerous other actions that can be taken to increase the likelihood that a program can be successfully executed once that business case is established. These include ensuring cost estimates are complete, accurate and updated regularly, and holding suppliers accountable through such activities as regular supplier audits and performance evaluations of quality and delivery. Moreover, we have recommended using metrics and controls throughout the life cycle to gauge when the requisite level of knowledge has been attained and when to direct decision makers to consider criteria before advancing a program to the next level and making additional investments.

NASA's Life Cycle for Flight Systems

NASA's life cycle for flight systems is defined by two phases—formulation[7] and implementation[8]—and several key decision points. These phases are then further divided into incremental pieces: Phase A through Phase F. See figure 1 for a depiction of NASA's life cycle for flight systems.

Project formulation consists of Phases A and B, during which time the projects develop and define requirements and the cost/schedule basis and design for implementation, including developing an acquisition strategy. During the end of the formulation phase, leading up to the preliminary design review (PDR),[9] the project team completes its preliminary design and technology development. *NASA Interim Directive 7120-97 for NASA Procedural Requirements 7120.5D, NASA Space Flight Program and Project Management Requirements*, specifies that during formulation, the project must complete a formulation agreement to establish the technical and acquisition work that must be conducted during this phase and define the schedule and funding requirements for that work. The formulation agreement has to describe activities, risk mitigation plans, and key tests to ensure that technologies will work as intended in a relevant environment, such as test chambers that simulate the conditions of space, by PDR, and the project is required to prepare, and continuously update, a technology readiness assessment. The project must also develop and document an agency baseline commitment[10] that includes the life-cycle cost estimate and other parameters. The formulation phase culminates in a review at key decision point C, known as project confirmation, where cost and schedule baselines are confirmed. Project progress can subsequently be measured against these baselines.

After a project is confirmed, it begins implementation, consisting of phases C, D, E, and F. Senior NASA officials must approve the project before it can proceed from one phase of implementation to another. A second design review, the critical design review (CDR),[11] is held during the latter half of phase C in order to determine if the design is stable enough to support proceeding with the final design and fabrication. After CDR and just prior to beginning phase D, the project completes a system integration review (SIR)[12] to evaluate the readiness of the project and associated supporting infrastructure to begin system assembly, integration, and test. In phase D, the project performs system assembly, integration, test, and launch activities. Phases E and F consist of operations and sustainment and project closeout.

NASA Projects Reviewed in GAO's Annual Assessment

NASA's mission is to drive advances in science, technology, and exploration to enhance knowledge, education, innovation, economic vitality, and stewardship of the Earth. NASA establishes many programs and projects that rely on complex instruments and spacecraft in order to accomplish its mission. NASA's portfolio of major projects ranges from robotic probes designed to explore the Martian surface, to satellites equipped with advanced sensors to study the earth, to telescopes intended to explore the universe, and spacecraft to transport humans and cargo beyond low-Earth orbit. In many cases, NASA's projects are expected to incorporate new and sophisticated technologies that must operate in harsh, distant environments.

This year, we assessed 21 major projects—6 projects in formulation and 15 projects in implementation. Five of the 15 projects in implementation successfully launched during 2011. The year after a project launches, we no longer include a 2-page summary in our annual report. However, we do maintain and continually assess historical cost, schedule, and performance information collected from these projects during the course of our reviews. When NASA determines that a project will have a life-cycle cost estimate of more than $250 million, we include that project in the next review. See table 1 for a list of the projects we reviewed in this year's assessment, and appendix III for a list of projects that we have reviewed from 2009 to 2012.

Table 1. 21 Selected Major NASA Projects Reviewed in GAO's 2012 Annual Assessment

Projects in formulation	ExoMars Trace Gas Orbiter (EMTGO)[a]
	Ice, Cloud, and Land Elevation Satellite-2 (ICESat-2)
	Orion Multi-Purpose Crew Vehicle (MPCV)
	Soil Moisture Active and Passive (SMAP)
	Solar Probe Plus (SPP)
	Space Launch System (SLS)
Projects in implementation	Aquarius[*]
	Global Precipitation Measurement (GPM) Mission
	Gravity Recovery and Interior Laboratory (GRAIL)[*]
	James Webb Space Telescope (JWST)
	Juno[*]
	Landsat Data Continuity Mission (LDCM)
	Lunar Atmosphere and Dust Environment Explorer (LADEE)
	Magnetospheric Multiscale (MMS)
	Mars Atmosphere and Volatile EvolutioN (MAVEN)
	Mars Science Laboratory (MSL)[*]
	NPOESS Preparatory Project (NPP)[*b]
	Orbiting Carbon Observatory 2 (OCO-2)
	Radiation Belt Storm Probes (RBSP)
	Stratospheric Observatory for Infrared Astronomy (SOFIA)
	Tracking and Data Relay Satellite (TDRS) Replenishment K and L[c]

Source: GAO analysis of NASA data.

[*] NASA projects that launched in 2011.

[a] In February 2012, NASA proposed canceling the EMTGO project as part of its fiscal year 2013 budget request.

[b] In January 2012, NASA announced that NPP had been renamed to Suomi National Polar-orbiting Partnership.

[c] TDRS Replenishment includes TDRS-K and TDRS-L. These satellites will launch separately, but are counted as only one mission.

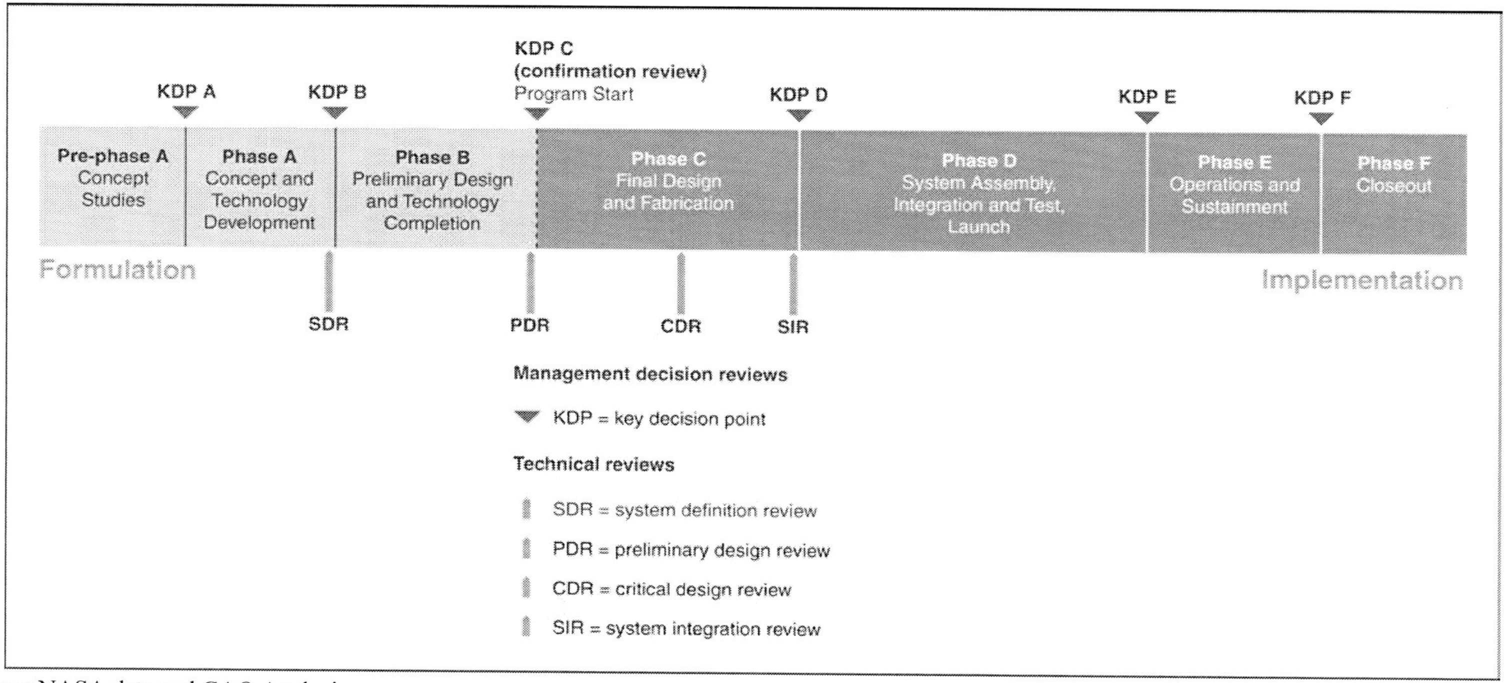

Source: NASA data and GAO Analysis.

Figure 1. NASA's Life Cycle for Flight Systems.

The portfolio of projects we reviewed includes a wide range of life-cycle costs. Life-cycle costs for projects in implementation range from $262.9 million for LADEE[13] to $8.8 billion for JWST. Life-cycle costs of projects in formulation have yet to be finalized, but current estimates range from $686 to $776 million for ICESat-2 to $18 billion through 2017 for SLS and Orion MPCV. Figure 2 depicts the launch date and life-cycle costs for projects we reviewed that are in the implementation phase.

NASA's Acquisition Management Remains on GAO's High Risk List

GAO has designated NASA's acquisition management as a high risk area since 1990 in view of persistent cost growth and schedule slippage in the majority of its major projects.[14] GAO's work continues to find that NASA has difficulty meeting cost, schedule, and performance goals for many of its projects. For example, over the past 3 years GAO has reviewed 13 projects with baselines established prior to 2009 that experienced an average development cost growth of almost 55 percent, with a total increase in development costs of almost $2.5 billion from their baselines established at confirmation. Several factors account for this cost and schedule growth, including poor cost estimating and underestimating risks associated with the development of its major systems.

In 2007, NASA developed a corrective action plan to improve how it manages its acquisitions. The plan identifies specific actions to strengthen project management, increase accuracy in cost estimating, facilitate monitoring of contractor cost performance, and improve business processes and financial management; the plan also establishes points of accountability and metrics to assess progress.

NASA has made some progress in the management and oversight of its major projects, which we have reported on in prior annual assessments and in our high risk report. For example, in 2005 we reported that NASA's acquisition policies did not conform to best practices for product development because they lacked major decision reviews at several key points in the project life cycle, which would allow decision-makers to make informed decisions about whether a project should be authorized to proceed in the development life cycle.[15] Based, in part, on our recommendations, NASA issued a revised policy that instituted several key decision points in the development life cycle for space flight programs and projects.

Furthermore, in 2011 we recommended that (1) NASA provide more transparency into project costs in the early phases of development so that Congress has sufficient information to conduct oversight and ensure earlier accountability, and (2) NASA develop a common set of measurable and proven criteria to assess design stability and amend its systems engineering policy accordingly.[16]

NASA agreed with the need to provide clarity on life-cycle cost range estimates and began providing increased information for projects in early formulation in its fiscal year 2013 budget submission. In addition, NASA is committed to enhancing its ability to monitor and assess the stability of programs and projects and is working to include additional criteria in its policy documents. We will continue to work with NASA as it makes progress toward more effectively assessing projects at key junctures in the project life cycle.

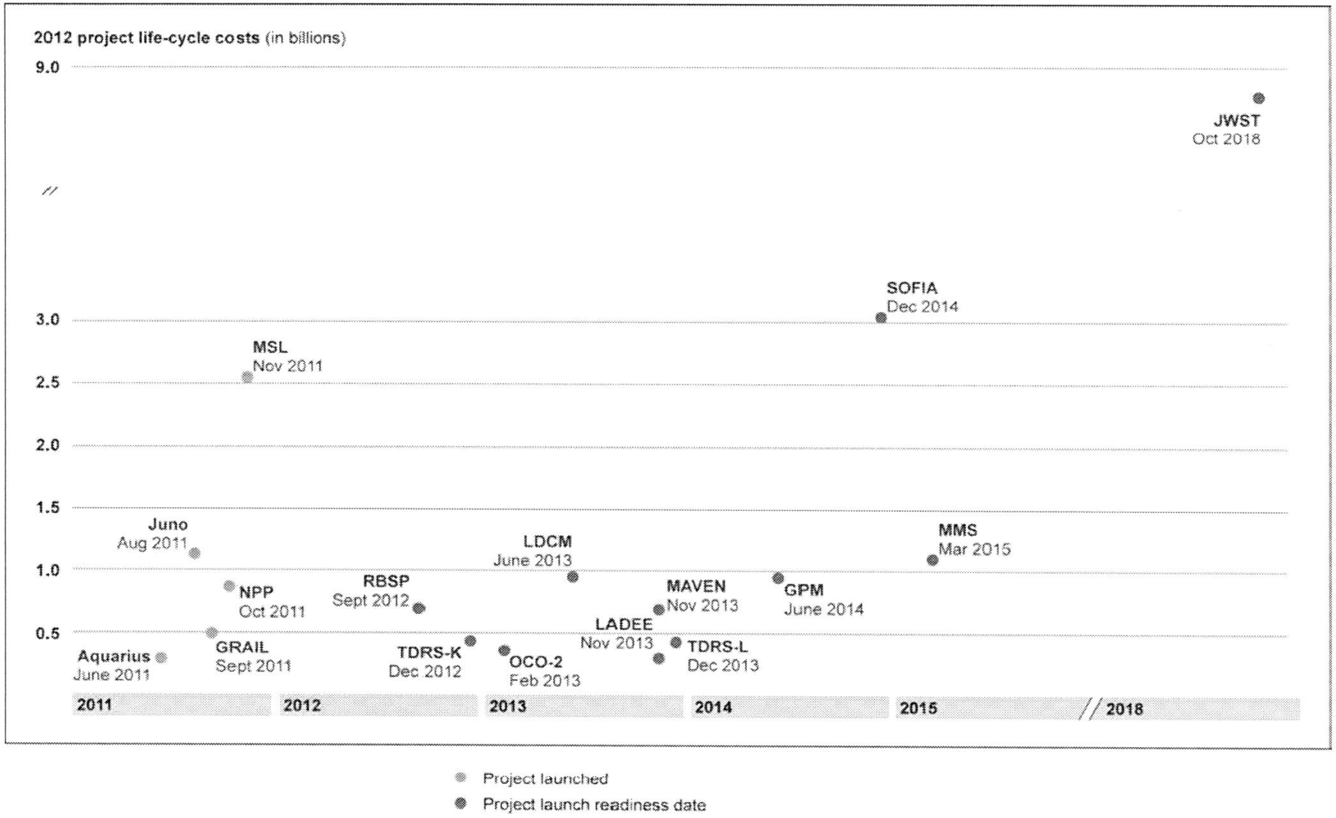

Source: GAO analysis of NASA data.

Note: TDRS-K and TDRS-L are planned to launch separately, but are counted as only one mission. NASA's total life-cycle cost for TDRS-K and TDRS-L is $425.5 million. Because launch is not a milestone that is applicable to SOFIA, we listed the date when the project is scheduled to achieve full operating capability. In January 2012, NASA reported that OCO-2's cost and schedule growth were under review because of a possible change in its launch vehicle.

Figure 2. Launch Date and Life-Cycle Cost for 15 Major NASA Projects in Implementation.

OBSERVATIONS ON NASA'S PORTFOLIO OF MAJOR PROJECTS

JWST Experienced Significant Cost and Schedule Growth, but Most Projects in Portfolio Are Currently Relatively Stable

Of the 15 projects in implementation,[17] the James Webb Space Telescope (JWST) has reported the most significant cost and schedule changes since last year. The development cost and schedule growth reported for JWST is not typical of the cost and schedule changes NASA has reported for its other major projects this year. Specifically, the JWST project has had over $3.6 billion—or 140 percent—in development cost growth and a schedule delay of over 4 years. To put the JWST project's development cost growth into perspective, its cost increase is over $443 million greater than the total life-cycle cost of the seven smallest major projects included in our review. See figure 3 below.

The impact of JWST's increases on the average development cost and schedule growth from project baselines is significant. For example, the 14 projects currently in implementation, excluding JWST, had an average development cost growth of $79 million—or 14.6 percent—and schedule growth of 8 months from their baselines. With JWST included, these numbers increase dramatically to almost 47 percent and 11 months respectively.

For most projects in implementation, development costs remained relatively stable or declined and launch dates did not slip in the past year. The LDCM, MMS, NPP, and TDRS Replenishment projects reported decreases in their development costs which resulted in a decline in each project's life-cycle cost. For example, NPP development costs declined because the project no longer needed $35 million in contingency funding to reach a February 2012 launch date since the project successfully launched in October 2011. Although development costs for the Aquarius, GRAIL, Juno, and RBSP projects declined, this decrease was fully or partially offset by increases to the projects' operations costs. The MSL project also transferred some funding from development to operations, and NASA reported a further increase in its life-cycle cost by providing an additional $59 million to operations to perform development activities during its cruise to Mars. See table 2 below for more details on projects' development cost and schedule growth reported this year and against their baselines.

Six of the 15 projects currently in implementation experienced significant development cost and/or schedule growth from their baselines.[18] Four of the 6 projects—NPP, SOFIA, Aquarius, and MSL—were highlighted in last year's report and account for 26 percent of the development cost growth and about 60 percent of the schedule growth in the portfolio. Two projects—JWST and GPM—experienced significant development cost and/or schedule growth from their baselines in the last year. The JWST project exceeded its development cost and schedule baselines by 140 percent and 52 months, respectively. These increases represent about 76 percent of the portfolio's development cost growth and approximately 30 percent of its schedule growth.

To the project's credit, NASA officials recently reported that all schedule milestones planned for 2011 had been accomplished. According to the project manager, GPM's launch has been delayed 11 months due to late deliveries of two primary instruments and spacecraft components by its contractors and the Japan Aerospace Exploration Agency (JAXA), but its development cost has actually declined because one instrument that was supposed to fly on a second spacecraft was removed from the project.

Table 2. Development Cost and Schedule Growth of Selected Major NASA Projects Currently in the Implementation Phase

Dollars in millions					
Project	**Cumulative development cost growth**	**Percentage cost growth**	**Development cost growth reported in past year**	**Cumulative launch delay (months)**	**Launch delay reported in past year (months)**
Aquarius	$33.2	17.2%	-$1.4	23	0
GPM[a]	-$35.9	-6.5%	$4.5	11	11
GRAIL[b]	-$28.7	-6.7%	-$28.7	0	0
Juno[c]	-$31.8	-4.3%	-$31.8	0	0
LADEE	$7.6	4.5%	$7.6	0	0
LDCM[d]	-$6.2	-1.1%	-$10.4	0	0
MAVEN	$0.0	0.0%	$0.0	0	0
MMS	-$0.1	0.0%	-$0.1	0	0
MSL[e]	$812.8	83.9%	-$20.6	26	0
NPP	$174.9	29.5%	-$12.2	42	0
OCO-2[f]	$0.0	0.0%	$0.0	0	0
RBSP[c]	-$3.0	-0.6%	-$3.1	4	4
SOFIA	$208.9	22.7%	$0.0	12	0
TDRS Replenishment[g]	-$25.8	-12.3%	-$8.6	0	0
Portfolio Excluding JWST					
Average	**$79.0**	**14.6%**	**-$7.5**	**8**	**1**
Portfolio Including JWST					
JWST	$3,616.8	140.1%	$3,487.0	52	52
Average	**$314.8**	**46.5%**	**$225.5**	**11**	**4**

Source: GAO analysis of NASA data.

Note: Shading indicates projects that exceeded the cost and/or schedule thresholds that trigger reporting to Congress under the law.

[a] GPM's development cost decreased because one instrument was removed from the project.

[b] The decline in GRAIL's development costs was partially offset by an increase in the project's operations costs.

[c] The life-cycle cost for the Juno and RBSP projects have not changed. The reduction in development costs was offset by an increase in operations costs.

[d] NASA reported LDCM's development costs were reduced because of progress in delivering an instrument and integrating it with the spacecraft and in completing environmental testing on another instrument.

[e] MSL established a new baseline in fiscal year 2010 after being reauthorized by Congress. MSL's development cost declined due to a reduction in the cost of its Atlas V launch vehicle and an estimated transfer in funding from development to operations.

[f] OCO-2's cost and schedule growth reported above reflect data as of October 2011. In January 2012, NASA reported that the project's cost and schedule were under review because of a possible change in its launch vehicle.

[g] NASA's development cost for TDRS Replenishment declined for several reasons, including greater than expected contributions from one of NASA's partners and the inadvertent inclusion of costs for another TDRS satellite.

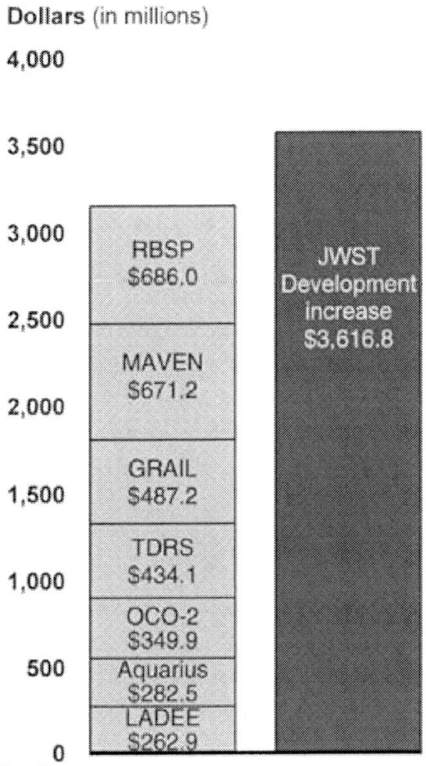

Source: GAO analysis of NASA data.

Figure 3. JWST's Cumulative Development Cost Growth Compared to the Total Life-Cycle Cost Estimates for Seven Selected Major NASA Projects.

The remaining nine projects entered implementation and established baselines in fiscal year 2009 or later and have not reported significant cost and schedule growth. For example, GRAIL and Juno successfully launched in 2011 and both projects were completed on time and within their allotted budget. Several of these projects, however, are entering, or have recently entered, the test and integration phase where cost and schedule growth is typically realized. Notably, some of the projects that are entering or have recently entered the test and integration phase have experienced similar challenges as some of the older projects that have reported cost and/or schedule growth, such as issues with maturing technology and/or not meeting design criteria. Although many of these projects have reported little or no cost growth from their baselines, several have experienced an increase in the value of their major contracts.[19] We plan to study this issue more going forward to determine why these values are changing and the extent to which changes in contract value could impact cost and schedule baselines.

Largest Space Missions Pose Portfolio Management Challenges to NASA

Although cost and schedule growth can occur on any NASA project, cost and schedule increases within NASA's most technologically advanced and costly projects can have cascading effects on the rest of its portfolio, presenting challenges to NASA management. For example, according to NASA officials, JWST's significant cost growth may lead to the postponement and possible cancellation of other science projects; and MSL's substantial cost overruns led NASA to take funding from other projects. JWST and MSL are two of NASA's largest projects and account for approximately 51 percent, or $11.4 billion, of the total life-cycle costs for projects in implementation during our review.

In addition, 2 of the 6 programs in formulation—SLS and Orion MPCV—are complex human spaceflight programs and together they are currently estimated to cost approximately $3 billion a year through 2017, representing about 17 percent of NASA's fiscal year 2012 budget. These programs have just recently transitioned from the Constellation program and, according to NASA officials, will not have measurable baselines established until February 2013.

In 2011, NASA and the Aerospace Corporation jointly studied typical agency missions. The study identified weaknesses in NASA's management of its projects and made several recommendations. For example, the study found that some projects had budget profiles that were unrealistic for the required effort.

The study recommended that a mission should not be initiated until an adequate funding profile is in place. Doing so is a key part of establishing a sound business case for a project, and best practices show matching resources to requirements increases the likelihood of program success. The study also found that in the past, some projects were confirmed with immature technology and design baselines and that committing to such baselines can lead to breaches and subsequent baseline changes. The study recommended that missions have a greater demonstrated readiness, which may lead to longer formulation periods with adequate funding for design and technology development. Although NASA's largest and most costly missions are going to be covered in a separate NASA study, the study concluded that these two recommendations extend to larger missions as well. These recommendations align with those that we have made in the past and are especially important for NASA's most technologically advanced and costly projects where problems can have cascading effects on the portfolio.

Observations from Our Assessment of Knowledge Attained by Key Junctures in the Acquisition Process

Many of NASA's projects are one-time articles, meaning that there is little opportunity to apply knowledge gained to the production of a second, third, or future increments of spacecraft. While space development programs are complex and difficult by nature and most are one-time efforts, NASA is still responsible for achieving what it promises when requesting and receiving funds.

We have previously reported that NASA would benefit from a more disciplined, knowledge-based approach to its acquisitions. For the projects reviewed this year, we continue to identify projects that have not met best practice standards for technology maturity and design stability and have experienced challenges in development. These challenges were assessed based on knowledge that, according to acquisition best practices, should be attained at key junctures in the project's life cycle to lessen the risks to the project.

Technology Challenges

Nearly two thirds of the projects in our current review do not meet best practice standards for technology maturity, but NASA has improved in this area in recent years. Our best practices work has shown that a technology readiness level (TRL) of 6—demonstrating a technology as a fully integrated prototype in a relevant environment—is the level of technology maturity that can minimize risks for space systems entering product development.[20] For NASA, a project enters product development—or implementation—following the project's preliminary design review and confirmation review. NASA's systems engineering policy states that by the preliminary design review a TRL of 6 is desirable prior to integrating a new technology in a project.[21] Demonstrating that technologies will work as intended in a relevant environment is a fundamental element of a sound business case, and its absence, we have found, is a marker for subsequent technical problems that must be addressed at the same time the system is being designed, fabricated, or tested.

Projects that are experiencing or have experienced technology challenges

- GPM
- GRAIL
- Juno
- JWST
- LADEE
- LDCM
- MMS
- MSL
- NPP
- SMAP
- SOFIA
- TDRS Replenishment

Specifically, 16 of the 21 projects in our review completed their preliminary design reviews. Of these 16 projects, 12 experienced technology challenges, including 10 projects that reported moving forward with immature technologies. See figure 4 for an analysis of the projects that we reviewed in the past 3 years that held their preliminary design review and the percentage of those projects that moved into implementation with immature technologies.

In our 2010 assessment of major NASA projects, 29 percent of the projects in our review had matured their technologies by the preliminary design review, and this figure rose to 38 percent last year. In this year's review, we found that the percent of projects meeting this criteria remained the same as last year.[22]

Our work has also shown that the use of heritage technology—proven components that are being modified to meet new requirements—can also increase the risk of problems when the items are not sufficiently matured to meet form, fit, and function standards of the project that will be using the component by the preliminary design review. Projects will modify the form, fit, and function of a heritage technology to adapt it to the new environment. For example, the size or weight of the component may change or the technology may function differently than its use on a previous mission. NASA frequently employs heritage technologies that have to be modified from their original form, fit, and function. NASA's *Systems Engineering Handbook* states that a frequently overlooked area is modification of heritage systems that are incorporated into different architectures, and operating in environments different from those in which they were designed to operate. Further, the *Handbook* states that project management tends to overestimate the maturity and applicability of heritage technology to a new project. Our work has shown, and NASA's own guidance concurs, that this is an area that is frequently underestimated when developing project cost estimates. Although NASA distinguishes critical technologies from heritage technologies, our work has shown critical technologies to be those that are required for the project to successfully meet customer requirements, regardless of whether or not the technologies are based on existing or heritage technology. Therefore, whether technologies are labeled as "critical" or "heritage," if they are important to the development of the spacecraft or instrument—enabling it to move forward in the development process—they should be matured by the preliminary design review.

Finally, there has been a decline in the amount of critical technology development reported for major NASA projects in our review. Specifically, we found that the average number of reported critical technologies per project declined from 4.7 in 2009 to 2.6 in 2012. See figure 5.

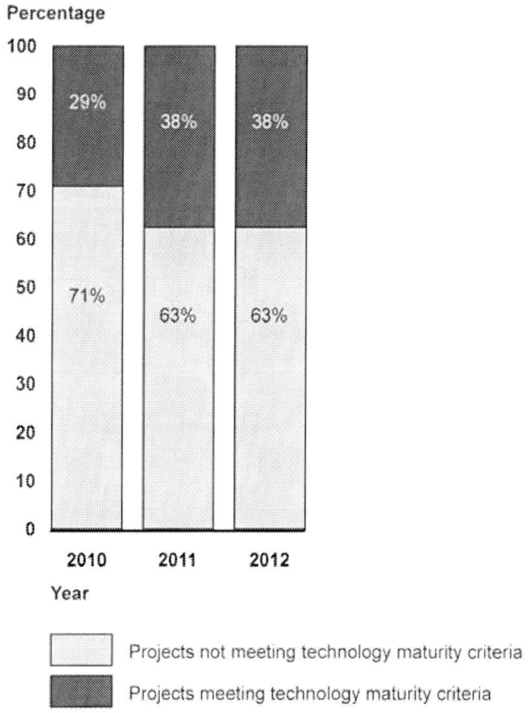

Source: GAO analysis of NASA data.

Note: The number of projects reviewed in each year has varied. See Appendix III. Additionally, totals may not add to 100 percent due to rounding.

Figure 4. Percentage of Selected Major NASA Projects Meeting and Not Meeting Technology Maturity Criteria at Preliminary Design Review.

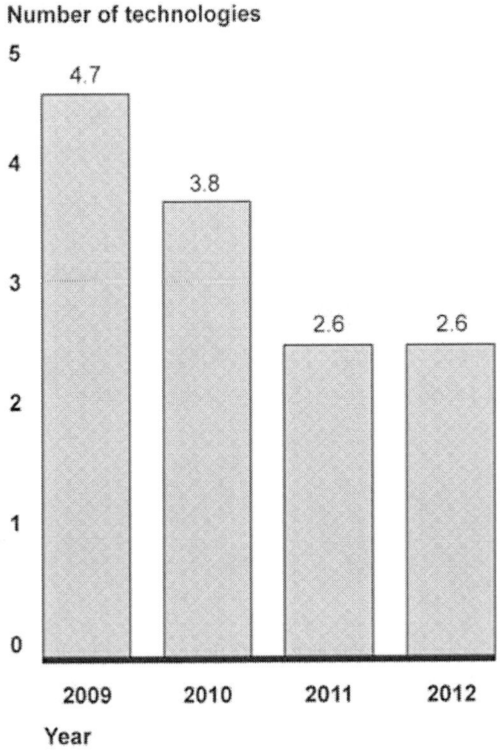

Source: GAO analysis of NASA data.

Note: Figures are based on the sum of reported critical technologies each year by projects in our review that had reached the preliminary design review. Each sum is then divided by the number of projects.

Figure 5. Average Number of Critical Technologies Reported for Selected Major NASA Projects Reviewed from 2009 to 2012.

Furthermore, we found the percentage of projects reporting one or no critical technologies at the preliminary design review has increased each year from 47 percent in 2009 to 65 percent in 2012. In last year's assessment, senior NASA officials stated that it appeared that the projects were not accurately identifying to us the number of critical technologies they planned to develop. They said that it appeared that projects had only identified critical technologies at the instrument—or system—level, and not the subsystem level. The data that NASA provided this year is comparable to what was submitted last year. We will continue to work with NASA to ensure that projects are accurately identifying their critical technologies to assist NASA decision makers in assessing the readiness of projects to move forward in their development life cycles.

Design Challenges

Thirteen[23] of the 14 projects in this year's review that had held a critical design review did not meet the best practices metric of having 90 percent engineering drawings in a releasable state.[24] The 14 projects averaged having only 62 percent of their engineering drawings releasable at their critical design reviews, the same percentage as we reported last year. See figure 6.

Projects that are experiencing or have experienced design challenges

- Aquarius
- EMTGO
- GPM
- JWST
- Juno
- LADEE
- MAVEN
- MMS
- MSL
- NPP
- OCO-2
- Orion-MPCV
- RBSP
- SOFIA
- SLS

The Tracking and Data Relay Satellite (TDRS) Replenishment project had met this criteria as of last year's review, but has had an increase in design drawings that dropped the project below the design stability criteria. The one project that did meet the criteria, the Orbiting Carbon Observatory 2 (OCO-2), is a rebuild of a prior design that launched and therefore a majority of its drawings were releasable. The performance of this year's portfolio against our design stability criteria continues a trend from prior assessments as nearly all projects we have reviewed since 2009 have failed to meet the criteria.

Our work that identified product development best practices shows that at least 90 percent of engineering drawings should be releasable by the critical design review. Guidance in NASA's *Systems Engineering Handbook* mirrors this metric. Previous discussions with project officials indicated the metric has been used inconsistently to gauge design stability. For example, Goddard Space Flight Center requires greater than 80 percent drawings released at the critical design review, yet in previous years several project officials reported that the "rule of thumb" for NASA projects is between 70 and 90 percent. As shown in figure 6 above, 7 of the 14 projects reported releasable engineering drawings of less than 70 percent, lower than the "rule of thumb" used by several project managers.

Because the critical design review is the time in a project's life cycle when the integrity of the project design and its ability to meet mission requirements is assessed, it is important that a project's design is stable enough to warrant continuing with the final design and fabrication phase. A stable design allows projects to "freeze" the design and minimize changes prior to beginning the fabrication of hardware, after which time re-engineering and re-work efforts due to design changes can be costly to the project in terms of time and funding. Some of the projects we reviewed in the past pointed to other activities that occurred prior to the critical design review as evidence of design stability. In addition to releasable engineering drawings, NASA often relies on subject matter experts in the design review process and other methods to assess design stability. Some projects also reported using engineering models and engineering test units to assess design stability, which in at least one case appears to be a successful approach. For example, despite having only 39 percent of its engineering drawings releasable which would indicate an unstable design, Juno project officials said that they were able to budget for and use engineering models for all instruments at critical design review. Officials stated that the use of engineering models helps decrease risk of flight unit development, and projects that did not use engineering models indicated that they might have caught problems earlier had they used them. However, project officials have also stated that engineering models are expensive to employ and not all projects have the available funds necessary to utilize them.

An indicator of an unstable design is the degree to which projects' design drawings increase post-critical design review. Projects that held their critical design review prior to fiscal year 2009 have reported having a lower percentage of releasable drawings at that review and a larger increase in engineering drawings post-critical design review than projects that have held their critical design review since that time. Over the past 4 years, we have reviewed 20 projects that have held their critical design review. As shown in figure 7 below, the 9 projects that held their critical design review prior to fiscal year 2009 have had, on average, a 184 percent increase in engineering drawings after the critical design review after having, on average, only 41 percent of drawings releasable at that review. The remaining 11 projects that held their critical design review in fiscal year 2009 or later have had, on average, only a 9 percent increase in engineering drawings and 72 percent of drawings releasable at that review. Several of the 11 projects have recently held their critical design review, limiting the amount of time over which growth in expected drawings might take place. Nonetheless, these projects represent a positive trend for NASA as they have achieved a higher level of design stability according to best practices and the agency's systems engineering policy, and have generally incurred less development cost growth than prior projects.

We have previously reported that NASA's acquisition policy does not specify a metric to measure a project's design stability at the critical design review, and last year, we recommended that NASA develop a common set of measurable and proven criteria to assess design stability and to amend NASA's systems engineering policy to that effect.[25] In response, NASA provided three technical indicators that were focused on design maturity. The three technical indicators were (1) the percentage of actual mass margin versus planned mass margin, (2) the percentage of actual power margin versus planned power margin, and (3) the percentage of overdue project requests for action.[26] When asked to provide data to support the use of these metrics, NASA's Chief Engineer told us that such data were not available and that the senior engineers consulted in their development agreed that the metrics are good indicators of design maturity. According to officials in NASA's Office of the Chief Engineer, NASA began a pilot this year with projects that went through their critical design review to determine the effectiveness of these indicators. NASA has not provided GAO any information on the results of this pilot. According to officials, NASA's program management and systems engineering policies are being updated to require projects to track these metrics. In addition, NASA also provided us information on additional technical indicators that can be used by projects. While they do not plan to require the use of these indicators, officials indicated that they plan on updating the agency's program management and systems engineering handbooks to encourage projects to use them to track design maturity. GAO expects to continue its dialogue with NASA regarding design metrics.

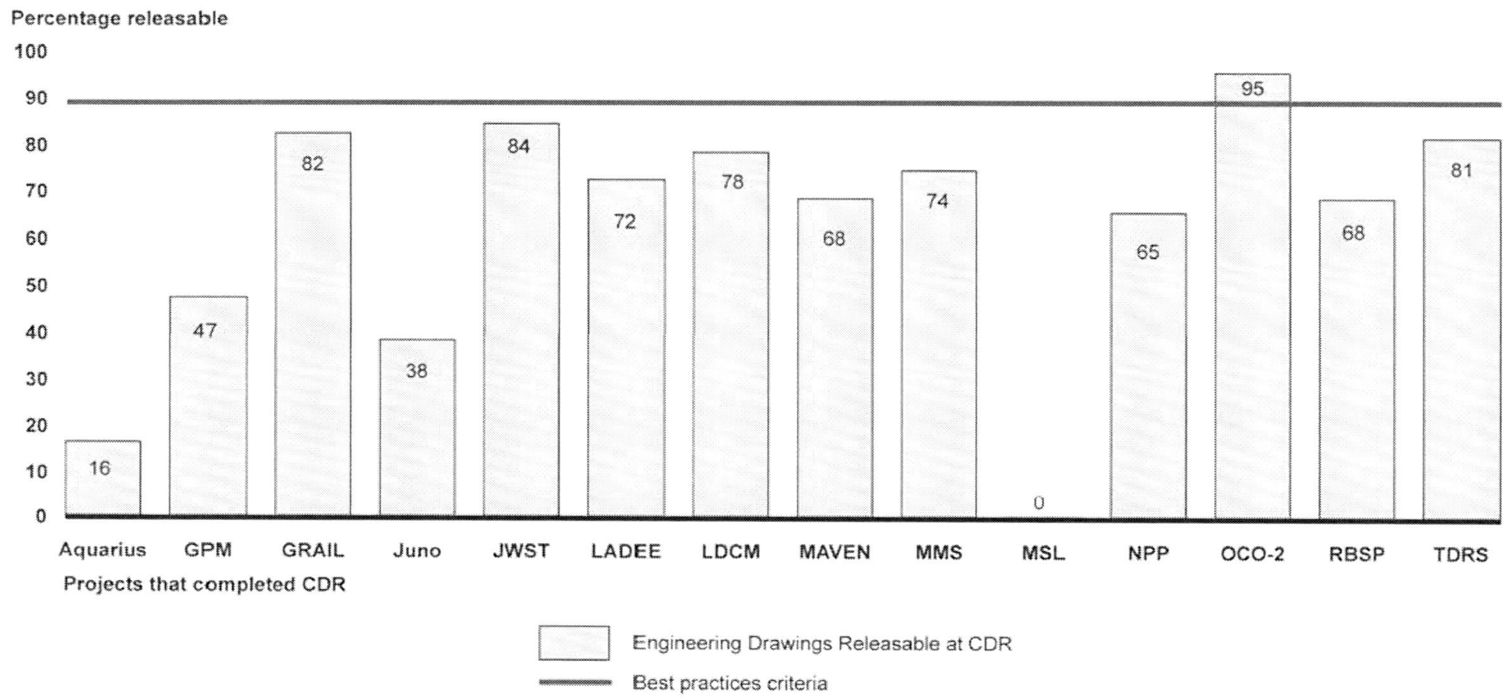

Source: GAO analysis of NASA data.

Figure 6. Percentage of Engineering Drawings Releasable at CDR for Selected Major NASA Projects.

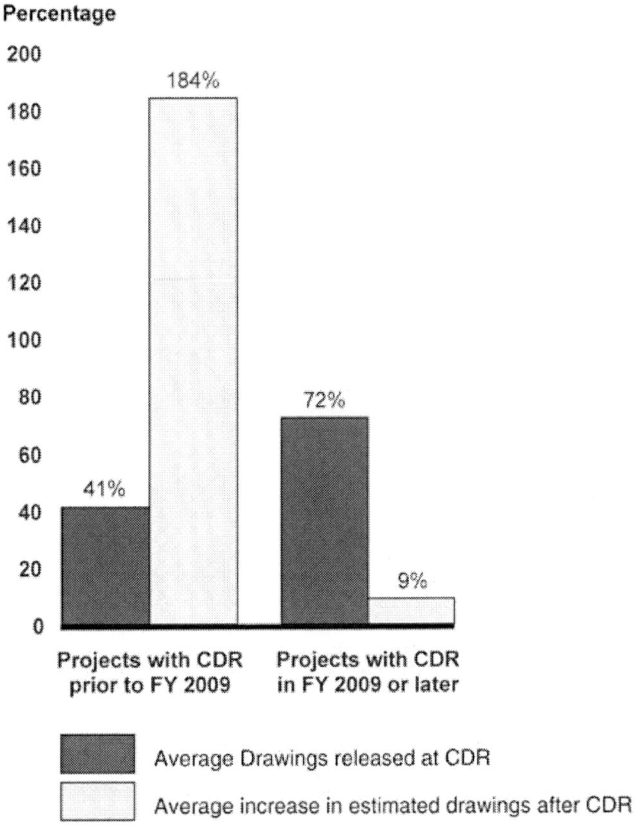

Source: GAO analysis of NASA data.

Figure 7. Comparison of Design Drawings Increase for 20 Major NASA Projects That Have Held CDR prior to and since Fiscal Year 2009.

Observations on Other Challenges That Can Affect Project Outcomes

In addition to collecting and analyzing data on the attainment of knowledge at key junctures, we identified five additional areas that can present challenges to obtaining positive project outcomes: launch vehicles, contractor management, parts, development partners, and funding.

Launch Vehicle Challenges

Nine of the 21 projects we reviewed reported challenges with launch vehicles, including the increasing cost and availability of launch vehicles.[27] Because the launch vehicle is integral to a project's design, the decision on which launch vehicle will be used should be made before preliminary design review. According to NASA officials, a delay or change in this decision can delay design or lead to costly re-design. For the last decade, NASA has relied principally on the Delta II medium class launch vehicle, which is built and sold by United Launch Alliance, to deliver science missions. The United States Air Force concluded its use of the Delta II launch vehicle in August 2009. The decreased demand for medium class launch vehicles, according to NASA, was insufficient to sustain the Delta II's production at prices traditionally paid; as a result, this vehicle is no longer a viable long-term option for its missions.[28] New medium class commercial launch vehicles are in development, but have not yet been certified—a process aimed at reducing risk that takes approximately 3 years. NASA could buy other mature launch vehicles for its missions, such as the Atlas V, but these vehicles would be considerably more expensive and provide excess capability than launch vehicles in the medium class. Another commercial launch vehicle, Taurus XL, has failed on its two most recent flights—the OCO and Glory spacecraft did not reach orbit—and will have to be recertified. Because of these challenges, some projects are facing uncertainties regarding which launch vehicle will be selected for the mission:

Projects that are experiencing or have experienced launch challenges

- GRAIL
- ICESat-2
- LADEE
- MAVEN
- NPP
- OCO-2
- SMAP
- SPP
- TDRS Replenishment

- The OCO-2 mission was scheduled to launch on the Taurus XL in February 2013. NASA, however, suspended OCO-2's Taurus XL task order pending a review of the launch vehicle's most recent failure. NASA officials have indicated that the OCO-2 project may use an alternate launch vehicle, but a decision has not yet been made. NASA officials have reported previously that changing the planned launch vehicle of a science mission after its preliminary design review is a fundamental change to the mission design and would lead to significant cost growth and schedule delays. Therefore, any change in the OCO-2 project's launch vehicle could have a significant impact on its cost and schedule. In January 2012, NASA officials stated that OCO-2's cost and schedule estimates were being reviewed pending a resolution of the launch vehicle issue.
- NASA officials told us they decided to use a Minotaur launch vehicle for at least one mission—LADEE—because of cost savings afforded by this vehicle and the lack of reliable and certified launch vehicles with the necessary performance. Minotaur vehicles are built from re-purposed Intercontinental Ballistic Missiles and use of these vehicles as a space transportation vehicle is subject to certain restrictions under federal law.[29] According to the law, NASA must certify to Congress that the Minotaur meets NASA's mission requirements, that the agency has the approval of

the Secretary of Defense, that the use is consistent with international obligations of the United States, and that doing so would provide cost-savings to the government—all of which, NASA officials stated, was done for the LADEE mission. Project officials told us they are also considering using a Minotaur for the OCO-2 and SMAP missions, but NASA has not yet determined if these projects meet the requirements of the law for using the Minotaur. Because of the lack of available alternative medium-class launch vehicles and the restrictions in using the Minotaur, a final decision on the launch vehicle for the SMAP mission has not been made and the project's preliminary design review, confirmation, and critical design review were delayed. According to project officials, they are currently basing the design of the project on using the Minotaur vehicle, but the project is at risk of costly redesigns if the vehicle ultimately selected is not the Minotaur. The project, however, is taking steps to minimize this risk in its design process.

- Project officials told us the launch vehicle for ICESat-2 remains undefined because of limited project funding and the high cost of suitable launch vehicles. The ICESat-2 mission is currently attempting to reach agreement with the Air Force to co-manifest with an Air Force spacecraft in order to reduce launch vehicle costs and stay within its life-cycle cost range estimate. Co-manifesting—or flying two or more spacecraft on a single launch vehicle—provides potential cost savings to the agencies involved. According to NASA officials, co-manifesting also comes with increased risk because it is difficult to coordinate the development and launch schedules, the orbit, and the destination of multiple missions. For example, a schedule delay on one project or mass increase on one spacecraft could adversely affect all missions involved. For the ICESat-2 mission, NASA is concerned that budget constraints may lead to a delay in the Air Force mission. If an agreement cannot be reached between the Air Force and NASA for co-manifesting the missions, project officials said they will work with NASA Launch Services to procure a launch vehicle to independently launch ICESat-2, which they added would be significantly more costly.

Contractor Management Challenges

Six of the 15 projects in implementation reported experiencing contractor challenges, including not completing work on time and inadequate communication that led to cost overruns and, in some cases, the need for increased oversight of contractors. Contractor performance is critical for the success of many NASA missions as NASA obligates about 85 percent of its annual budget on contracts.

The JWST project has encountered contractor management challenges with its prime contractor Northrop Grumman and Lockheed Martin, a major sub-contractor for the project's primary imaging instrument. In 2010, an independent review panel attributed some of the project's cost growth and schedule delays to program management issues. Specifically, the panel cited inadequate communication between NASA and its contractors and the absence of a project representative at the prime contractor facility. In response to the panel's report, NASA officials restructured the JWST project and directed it to assume the lead role over systems engineering functions and some of the integration and testing responsibility from Northrop Grumman, which the company agreed was the proper action. In addition, the project and its prime contractor are planning to have at least one senior JWST project representative reside at Northrop Grumman. A NASA official added that the representative will be staffed at the Northrop Grumman facility by the end of fiscal year 2012 and will serve as the day-to-day liaison with the contractor, providing quick responses to contractor questions and clarifying the project's directions. JWST officials also told us that the project encountered between $60 million and $200 million in cost overruns when Lockheed Martin was unable to deliver one of the scientific instruments on time. As a result, officials said they have staffed two project representatives at Lockheed Martin whose responsibilities include providing technical guidance on contractor questions and coordinating discussions between project and contractor staff to facilitate timely resolution of technical issues.

Projects that are experiencing or have experienced contractor management challenges

- GPM
- GRAIL
- JWST
- MMS
- RBSP
- TDRS Replenishment

The MMS and GPM projects also reported that contractors did not complete work on time. For example, the MMS project's prime contractor, the Southwest Research Institute, was having difficulty delivering one of the scientific instruments on time, and the project reported that recent detailed cost estimates indicate that the cost to complete its development will exceed the overall available budget by $15.9 million. In another example, Northrop Grumman did not have the capacity to produce sufficient quantities of flight power system electronic boards and command and data handling units, which affected the integration and test schedule for the GPM project. In response, the project had to use a second manufacturer to help mitigate this production issue.

MAVEN project officials reported that they proactively took steps early in development to mitigate potential contractor challenges. For example, the project received NASA approval to have a number of mission quality assurance staff reside at Lockheed Martin, one of the project's prime contractors. In addition, after finding deviations in a vendor's part used on a component that controls and stabilizes the spacecraft, officials said they instituted additional, mandatory inspections during manufacturing and assembly of the part.

Parts Challenges

Thirteen of the 15 projects in implementation reported parts, materials, and process issues, some of which resulted in cost growth and/or schedule delays. The cost impacts generally ranged from tens of thousands to several million dollars during a project's development phase. One project—NPP—experienced cost increases of over $100 million due to parts problems with partner-provided instruments. In some cases, the problems also led to instruments being delivered late. These findings are consistent with our June 2011 report, in which we reported that parts quality problems have endangered entire missions for space and missile defense acquisitions, especially when those problems were discovered late in the development cycle.[30] According to NASA officials, parts problems are not uncommon for projects and NASA's testing process is designed to identify parts failures at the component, subsystem, and system level before they lead to mission failure. Below are some examples of the parts problems encountered by projects in this year's assessment:

Projects that are experiencing or have experienced parts challenges

- Aquarius
- GPM
- GRAIL
- JUNO
- JWST
- LDCM
- MMS
- MSL
- NPP
- OCO-2
- RBSP
- Sofia
- TDRS Replenishment

- The RBSP and MMS projects experienced problems with the same part, a high-voltage optocoupler. The problem cost the RBSP project about $900,000 and required the manufacturer to make revisions to the part's design. Issues with the part also contributed to a 2-month delay in delivering one of the MMS project's scientific instruments at a cost impact of approximately $500,000.
- The LDCM project experienced a 9-month delay in the delivery of the spacecraft's star tracker when a stud broke during assembly that held key components. The contractor had developed a new design process that was supposed to improve the star tracker's detector, but it failed to do so. Consequently, the contractor reverted to its original design process.
- The LDCM and OCO-2 projects encountered problems with black chrome coating used to protect optically sensitive instruments. For the LDCM project, approximately $3 million and 2 months of project-held schedule reserve were used to address this issue. For the OCO-2 project, the cost impact was below $1 million and resulted in a schedule slip of about 2 months.

In June 2011, we recommended that NASA implement a mechanism for a periodic, governmentwide assessment and reporting on parts quality problems in major space and missile defense programs, with periodic reporting to Congress. The assessment would include the frequency such problems are appearing in major programs, changes in frequency from previous years, and the effectiveness of corrective measures. NASA concurred with our recommendation.

Development Partner Challenges

Four projects reported challenges with domestic or international partners' not meeting project commitments within planned resources. These challenges included lack of adequate partner funding, late delivery of partner instruments, and technical issues with spacecraft. Some of the problems were outside the development partners' control while others were more likely caused by the partners themselves.

Projects that are experiencing or have experienced development partner challenges

- Aquarius
- GPM
- LDCM
- MMS

Two projects reported budget concerns and setbacks in instrumentation development involving their domestic partners. LDCM project officials told us that NASA is concerned that the U.S. Geological Survey (USGS) may not be able to meet its funding commitments now and during ground operations. Project officials said that NASA has had to cover project development costs that were supposed to be paid for by USGS because USGS did not have an adequate budget to support the work. NASA made a corresponding decrease to its expected operations costs in order to keep its life-cycle costs stable with the understanding that USGS would have a sufficient budget during the operations phase to cover this decrease. USGS will manage the ground systems and is responsible for funding flight and mission operations, taking control of the project after launch. In another example, NPP project officials said they are concerned about the longevity of the partner-provided instruments on orbit because they were designed and built in undisciplined environmental conditions, resulting in numerous design and test failures.

Projects with international partners encountered schedule delays due to technical issues and, in one case, a natural disaster. For example, Argentina's National Committee of Space Activities was responsible for spacecraft development delays that resulted in a 23-month launch delay of the Aquarius project. Project officials said that while Argentina's National Committee of Space Activities is technically competent, it lacks experience in managing spacecraft production projects. In addition, Aquarius project officials reported that certain aspects of the spacecraft, such as power continuity, threaten Aquarius's performance during operations, but they deem this risk as very unlikely as NASA subject matter experts participated in the design reviews and development activities for the solar array. The development partner for the GPM project, JAXA, faced an unexpected setback when the March 2011 earthquake and tsunami hit Japan. GPM project officials reported that the delivery of the Dual-Precipitation Radar slipped by 4 months after the natural disaster struck Japan, where the instrument is being constructed.

Funding Challenges

Although more than half of the projects in our review experienced funding challenges, we did not find consistent reasons among the projects for these issues. Challenges experienced by a few projects, however, significantly affected the project's performance, the viability of NASA's broader portfolio, and have the potential to do so in the future. We view this challenge as an area that will require close attention in the next few years as NASA moves forward with large investments in its human spaceflight programs and the full effect of the cost increases for the JWST project becomes apparent.

Projects that are experiencing or have experienced funding challenges

- Aquarius
- EMTGO
- GPM
- ICESat-2
- JWST
- LDCM
- MMS
- MPCV
- OCO-2
- SLS
- SMAP
- SOFIA

In October 2011, NASA announced that the JWST project will cost $3.7 billion more than previously expected and take over 4 years longer to develop. The result of the cost increase on other projects in the portfolio has yet to be fully addressed by NASA, but the agency has indicated that the impacts being assessed would delay some future science missions planned for launch after 2015. In addition, NASA officials said that other projects, starting with those that have not yet been confirmed, may face possible cancellation to help offset JWST's cost growth. The impact of the JWST project's cost growth appears to be affecting at least one project in the portfolio as the Administration has proposed to terminate funding for the EMTGO project—part of the Mars Exploration Joint Initiative between NASA and the European Space Agency (ESA)—and planning for the NASA/ESA Mars 2018 mission concept in the fiscal year 2013 President's budget request due to funding constraints. Prior to this announcement, NASA officials stated that the agency delayed EMTGO's scheduled January 2012 confirmation due to a NASA review of the Mars program, which was initiated as a result of the agency's overall funding constraints. These constraints include the recent JWST project's cost increases. In addition, NASA had originally planned to provide an Atlas V launch vehicle for EMTGO. However, NASA officials report that in September 2011 the agency informed ESA that they should not expect NASA to be able to provide the launch vehicle for the 2016 mission due to budgetary reasons.

As NASA confronts JWST's cost growth and its repercussions, it will also be initiating two other large-scale investments in human spaceflight—the new SLS and the Orion MPCV—which together will require approximately $3 billion a year. Recent human space flight programs experienced significant cost growth—partly reflective of the technical and design risks in developing these systems, but also of poor management and oversight practices such as establishing funding profiles that fit annual budgets but do not match resources needed in the early phases of the development process. Moreover, these two programs face highly ambitious schedules that are likely to require the agency to invest more heavily than is currently anticipated. NASA officials stated that the agency conducted a comprehensive review of these programs for the President's fiscal year 2013 budget proposal, issued in February 2012. The agency, however, will not be able to provide a baseline life-cycle cost estimate for these programs until February 2013 when NASA expects to have greater clarity of the issues surrounding integration of the programs.

Observations about NASA's Continued Efforts to Improve Its Acquisition Management

NASA has been implementing initiatives to reduce acquisition management risk, which has been on GAO's high risk list for more than 20 years.[31] Specifically, NASA identified five areas for improvement—program/project management, cost reporting process, cost estimating and analysis, standard business processes, and management of financial management systems—each of which contains targets and goals to measure improvement. One of the most prominent efforts is the Joint Cost and Schedule Confidence Level (JCL), which is designed to help project officials with management, cost and schedule estimating, and maintenance of adequate levels of funding reserves. Another important effort is the implementation of earned value management (EVM) within certain programs and specific in-house efforts to help projects monitor the scheduled work done by contractors and employees. EVM is a program management tool being implemented at NASA centers that integrates technical, cost, and schedule parameters of a contract and uses those parameters to measure cost and schedule variances. This management tool, however, has not yet been institutionalized within the NASA centers. These efforts are positive steps toward addressing NASA's issues with meeting cost and schedule baselines; however, it is too early to assess the impact on NASA's performance.

Challenges Related to JCL Implementation and Oversight

NASA's policy requires that a JCL be developed prior to a project's confirmation review.[32] The JCL is a probabilistic analysis that includes, among other things, all cost and schedule elements, incorporates and quantifies potential risks, assesses the impacts of cost and schedule to date, and addresses available annual resources to arrive at development cost and schedule estimates associated with various confidence levels. The primary goals of the JCL are to provide assurance to stakeholders that NASA will meet cost and schedule targets and provide transparency on the effects of funding changes on the probability of meeting cost and schedule commitments. In general, projects' cost and schedule baselines are based on a 70 percent confidence level,[33] unless the decision authority approves a different confidence level with appropriate justification and documentation. The cost baseline includes unallocated future expenses, which refer to the portion of resources identified in JCL probabilistic calculations that cannot yet be allocated to specific sub-elements of a program or project's plan as potential risks have not yet been realized. The move to the JCL process, or probabilistic estimating, marks a major departure from NASA's prior practice of establishing a point estimate and adding a percentage on top of that point estimate to provide for contingencies. When we use the term reserves in this report, we are referring to the unallocated future expenses held at the Mission Directorate level or at the project level, both of which are identified through JCL probabilistic calculations. Finally, projects must also be funded at a level equivalent to at least the 50 percent confidence level.[34] The remaining funding is held as reserves at the Mission Directorate.

Five projects in our review—LADEE, LDCM, MAVEN, MMS and OCO-2—completed the JCL process according to NASA's policy.[35] NASA officials told us, however, that a few projects have excluded or not fully considered relevant cost inputs and risks, such as launch vehicle costs and risks associated with development partner challenges. For example, the OCO-2 project's JCL estimate did not include the Taurus XL launch vehicle costs. NASA officials have also noted that projects have varying levels of proficiency in preparing JCLs. The LADEE project's JCL, for example, shows a narrow range of approximately $2 million between the 50 and 70 percent confidence levels. According to NASA officials, this narrow difference is indicative of an incomplete consideration of possible risks. The LADEE project's Standing Review Board also noted that the tool used to prepare the project JCL models risk in such a way that underestimates uncertainty. The Standing Review Board utilized independent cost and schedule estimates to support its recommendation that NASA add $25 million in reserves and to push the launch readiness date back by a month or two. NASA officials emphasized that a primary goal moving forward is to be more consistent across projects in implementing the JCL policy. In 2009, NASA's Cost Analysis Division, established the development of a JCL implementation handbook as a key next step. According to NASA officials, they anticipate completing the handbook in fiscal year 2012.

We were unable to independently confirm that the five projects that prepared a JCL were budgeted at the 70 percent confidence level or at a different confidence level approved by the decision authority. NASA subsequently provided additional documentation to clarify this issue, but it was not received in time to be included in our analysis. It is important that projects are budgeted to the appropriate confidence level given past studies that have linked cost growth to insufficient reserves, poorly phased funding profiles, and more generally, optimistic estimating practices.

NASA has not yet launched a mission that used a JCL to support the establishment of the project baseline at the confirmation review. As a result, NASA officials stated it is too early to determine whether or not implementation of the JCL policy has helped current projects avoid cost and schedule overruns experienced by the agency's major projects in the past. NASA officials estimate that it will take up to 7 years to evaluate the impact and effectiveness of the JCL. We agree that it is too early to evaluate the policy until a sufficient number of projects have launched, permitting a valid comparison between estimated and actual costs. It is important to note, however, that the baselines for NASA's Juno and GRAIL projects were established using a prior cost estimating policy, which required the projects to be budgeted at the 70 percent confidence level for cost, and both projects met their baseline commitments.

Implementation of Earned Value Management at NASA Centers in Progress

NASA's goal is to develop and deploy an agencywide EVM capability that is compliant with generally accepted standards.[36] As discussed in GAO's *Cost Estimating and Assessment Guide*,[37] if implemented appropriately, EVM provides objective reports of project status, produces early warning signs of impending schedule delays and cost overruns, and can identify specific development efforts contributing to those overruns. While EVM is being used by some projects at NASA, it is not yet clear that it is being used consistently by the projects as a tool for managing cost and schedule. We have ongoing work assessing whether NASA's major projects are effectively using EVM techniques to manage their acquisitions and plan to issue a report in 2012.

PROJECT ASSESSMENTS

The 2-page assessments of the projects we reviewed provide a profile of each project and describe the challenges we identified this year as well as challenges that we have identified in the past. On the first page, the project profile presents a general description of the mission objectives for each of the projects; a picture of the spacecraft or aircraft; a schedule timeline identifying key dates for the project; a table identifying programmatic, launch, and contract information; a table showing the current baseline year cost and schedule estimates and the January 2012 cost and schedule data; a table showing the challenges relevant to the project; and a project summary narrative. To maintain information on challenges the projects experience over their lifetime, we continued to identify project challenges that were previously reported. On the second page of the assessment, we provide an analysis of the project challenges, and outline the extent to which each project faces cost, schedule, or performance risk because of these challenges, if applicable. NASA project offices were provided an opportunity to review drafts of the assessments prior to their inclusion in the final product, and the projects provided both technical corrections and more general comments. We integrated the technical corrections as appropriate and characterized the general comments below the project update.

See figure 8 below for an illustration of the layout of each 2-page assessment.

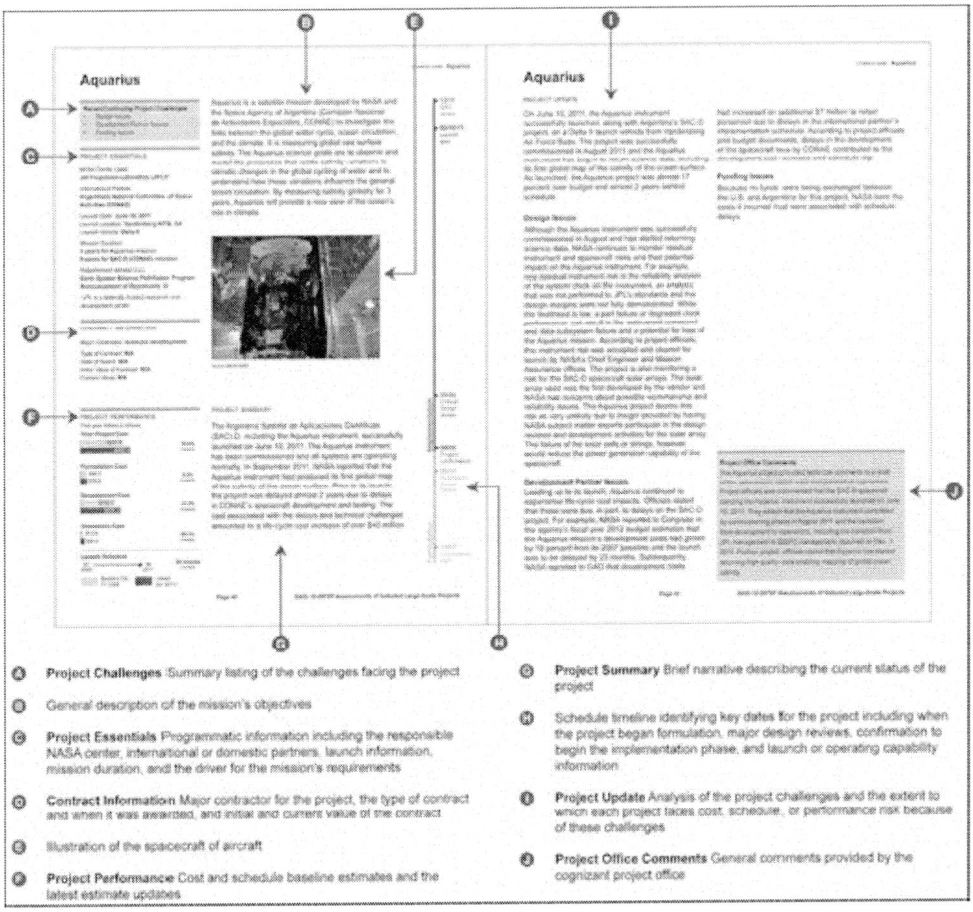

Source: GAO analysis.

Figure 8. Illustration of a Project's 2-Page Summary.

Aquarius

Recent/Continuing Project Challenges

- Design Issues
- Development Partner Issues
- Funding Issues

Project Essentials

NASA Center Lead: **Jet Propulsion Laboratory (JPL)***

International Partner: **Argentina's National Committee of Space Activities (CONAE)**

Launch Date: **June 10, 2011**

Launch Location: **Vandenberg AFB, CA**

Launch Vehicle: **Delta II**

* JPL is a federally funded research and development center.

Mission Duration:
3 years for Aquarius mission
5 years for SAC-D (CONAE) mission
Requirement derived from: **Earth System Science Pathfinder Program Announcement of Opportunity 3**

Contract Information

Major Contractor: **In-house development**
Type of Contract: **N/A**
Date of Award: **N/A**
Initial Value of Contract: **N/A**
Current Value: **N/A**

Project Performance

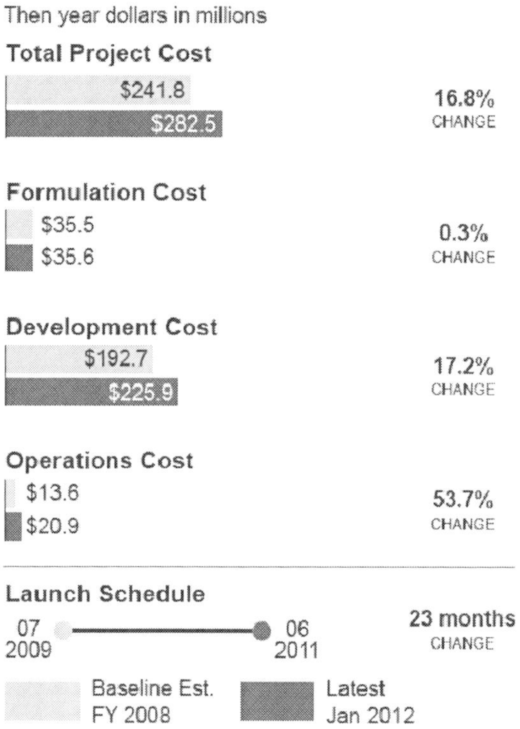

Aquarius is a satellite mission developed by NASA and the Space Agency of Argentina (Comisión Nacional de Actividades Espaciales, CONAE) to investigate the links between the global water cycle, ocean circulation, and the climate. It is measuring global sea surface salinity. The Aquarius science goals are to observe and model the processes that relate salinity variations to climatic changes in the global cycling of water and to understand how these variations influence the general ocean circulation. By measuring salinity globally for 3 years, Aquarius will provide a new view of the ocean's role in climate.

Source: NASA/VAFB.

Project Summary

The Argentine Satélite de Aplicaciones Científicas (SAC)-D, including the Aquarius instrument, successfully launched on June 10, 2011. The Aquarius instrument has been commissioned and all systems are operating normally. In September 2011, NASA reported that the Aquarius instrument had produced its first global map of the salinity of the ocean surface. Prior to its launch, the project was delayed almost 2 years due to delays in CONAE's spacecraft development and testing. The cost associated with the delays and technical challenges amounted to a life-cycle cost increase of over $40 million.

Project Update

On June 10, 2011, the Aquarius instrument successfully launched, along with Argentina's SAC-D project, on a Delta II launch vehicle from Vandenberg Air Force Base. The project was successfully commissioned in August 2011 and the Aquarius instrument has begun to return science data, including its first global map of the salinity of the ocean surface. As launched, the Aquarius project was almost 17 percent over budget and almost 2 years behind schedule.

Design Issues

Although the Aquarius instrument was successfully commissioned in August and has started returning science data, NASA continues to monitor residual instrument and spacecraft risks and their potential impact on the Aquarius instrument. For example, one residual instrument risk is the reliability analysis of the system clock on the instrument, an analysis that was not performed to JPL's standards and the design margins were not fully demonstrated. While the likelihood is low, a part failure or degraded clock performance can result in the instrument command and data subsystem failure and a potential for loss of the Aquarius mission. According to project officials, this instrument risk was accepted and cleared for launch by NASA's Chief Engineer and Mission Assurance offices. The project is also monitoring a risk for the SAC-D spacecraft solar arrays. The solar array used was the first developed by the vendor and NASA has concerns about possible workmanship and reliability issues. The Aquarius project deems this risk as very unlikely due to insight provided by having NASA subject matter experts participate in the design reviews and development activities for the solar array. The failure of the solar cells or strings, however, would reduce the power generation capability of the spacecraft.

Development Partner Issues

Leading up to its launch, Aquarius continued to experience life-cycle cost impacts. Officials stated that these were due, in part, to delays on the SAC-D project. For example, NASA reported to Congress in the agency's fiscal year 2012 budget estimates that the Aquarius mission's development costs had grown by 18 percent from its 2007 baseline and the launch was to be delayed by 23 months. Subsequently, NASA reported to GAO that development costs had increased an additional $7 million to retain personnel due to delays in the international partner's implementation schedule. According to project officials and budget documents, delays in the development of the spacecraft bus by CONAE contributed to the development cost increase and schedule slip.

Funding Issues

Because no funds were being exchanged between the U.S. and Argentina for this project, NASA bore the costs it incurred that were associated with schedule delays.

PROJECT OFFICE COMMENTS

The Aquarius project provided technical comments to a draft of this assessment, which were incorporated as appropriate. Project officials also commented that the SAC-D spacecraft carrying the Aquarius instrument successfully launched on June 10, 2011. They added that the Aquarius instrument completed its commissioning phase in August 2011 and the transition from development to operations, including the transition from JPL management to GSFC management, occurred on Dec. 1, 2011. Further, project officials stated that Aquarius has started returning high quality data enabling mapping of global ocean salinity.

ExoMars Trace Gas Orbiter

Recent Project Challenges

- Funding Issues

Project Essentials

NASA Center Lead: **Jet Propulsion Lab**
International Partner: **European Space Agency**
Projected Launch Date: **N/A**
Launch Location: **N/A**

Launch Vehicle: **N/A**

Mission Duration: **N/A**

Requirement derived from: **2011 Planetary Decadal Survey Source: ESA (artist depiction).**

Contract Information

Major Contractor: **The project is in formulation, and no prime contractor has yet been selected.**

Type of Contract: **N/A**

Date of Award: **N/A**

Initial Value of Contract: **N/A**

Current Value: **N/A**

Project Performance

Then year dollars in millions

Latest: Feb 2012 **Proposed for termination**

* The project has not yet reached the point in the acquisition life cycle where a preliminary life-cycle cost estimate would normally be developed.

Preliminary estimate of Project Life Cycle Cost*

Launch Schedule N/A

The ExoMars Trace Gas Orbiter (EMTGO) was planned to be the first of two joint missions that were planned to be developed by NASA and the European Space Agency (ESA) for launch in 2016 and 2018. EMTGO was envisioned to investigate trace gases on Mars that may be signatures of active biological and/or geographical processes. EMTGO instruments were to be designed to work across different bands of the electromagnetic spectrum, including infrared, visible, and ultraviolet light. EMTGO was to provide the data relay services for future missions.

Source: ESA (artist depiction).

Project Summary

The Administration has proposed to terminate funding for NASA's EMTGO project and planning for the NASA/ESA Mars 2018 mission concept in the Fiscal Year 2013 President's Budget Request due to budgetary constraints. Prior to this announcement, NASA was responsible for four EMTGO instruments that were competitively selected in August 2010, the Electra Relay Radio, and the Science Operations. The NASA hardware was largely based on heritage technology. ESA is currently planning to provide the spacecraft, one instrument, and an Entry Descent Module. ESA began development about two years ahead of NASA.

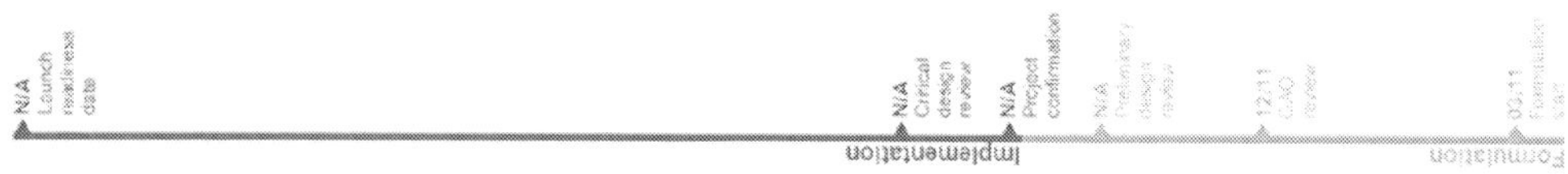

Project Update

EMTGO was part of Europe's long-term plan for the robotic and human exploration of the solar system. ESA was to provide the Trace Gas Orbiter, which would orbit Mars to search for trace gases, indicate the variation and location of them, and provide images of the surface related to where those gases emanate and where they are removed from the atmosphere. The orbiter was also to perform as a telecommunications link for this and potentially future missions. ESA was planning to provide the spacecraft, one instrument, and an Entry Descent Module to demonstrate the ESA's ability to land on Mars. NASA was to be responsible for four instruments on the Trace Gas Orbiter: the Mars Atmosphere Trace Molecule Occultation Spectrometer, the ExoMars Climate Sounder, the Mars Atmosphere Global Imaging Experiment, and the High Resolution Stereo Color Imager. NASA was also responsible for the Electra Relay Radio and the Science Operations for the mission. According to project officials, two of the instruments have flown previously, the other two are new designs with heritage elements, and the Electra Relay Radio is a copy of the radio for the MAVEN project. According to ESA, this technology demonstration would allow the agency to test key technologies for potential future missions.

Funding Issues

The Administration has proposed to terminate funding for NASA's EMTGO project and planning for the NASA/ESA Mars 2018 mission concept in the Fiscal Year 2013 President's Budget Request due to funding constraints. Prior to the announcement of the proposed cancellation, agency had delayed EMTGO's scheduled January 2012 confirmation due to a NASA review of the Mars program, which was initiated as a result of the agency's overall funding constraints.

These constraints include the recent JWST project's cost increases. NASA's Director of the Planetary Science Division testified to Congress in November 2011 that NASA and ESA were jointly reviewing the 2016 mission given increasing budget pressures. In addition, NASA had originally planned to provide an Atlas V launch vehicle for EMTGO. However, NASA officials report that in September 2011 the agency informed ESA that it should not expect NASA to be able to provide the Launch Vehicle for the 2016 mission due to budgetary reasons.

Project Office Comments

The EMTGO project provided technical comments to a draft of this assessment, which were incorporated as appropriate. NASA officials also commented that the agency has made tough but sustainable choices to provide stability and continuity to existing programs and set an affordable pace for unfolding the next great chapter in exploration. They added that NASA will not be moving forward with the planned 2016 and 2018 ExoMars missions that it had been exploring with the European Space Agency. Instead NASA will develop an integrated strategy to ensure that the next steps for robotic Mars Exploration program will support science as well as human exploration goals, technology advances, and to potentially take advantage of the 2018-2020 Mars exploration window. The officials reported that the budget provides support for this new approach, and this process will be informed by extensive coordination with the science community and our international partners.

Global Precipitation Measurement Mission

Recent/Continuing Project Challenges

- Funding Issues
- Development Partner Issues
- Contractor Issues

Previously Reported Challenges

- Technology Issues
- Design Issues

Project Essentials

NASA Center Lead: **Goddard Space Flight Center**
International Partner: **Japan Aerospace Exploration Agency (JAXA)**
Projected Launch Date: **June 2014**
Launch Location: **Tanegashima Island, Japan**
Launch Vehicle: **JAXA supplied**
Mission Duration: **3 years**
Requirement derived from: **Revalidated in the Earth Science Decadal Survey, 2007**

Contract Information

Major Contractor: **Ball Aerospace and Technologies Corp.**
Type of Contract: **Cost Plus Award Fee**
Date of Award: **2005**
Initial Value of Contract: **$97.6 million**
Current Value: **$214.6 million**

Project Performance

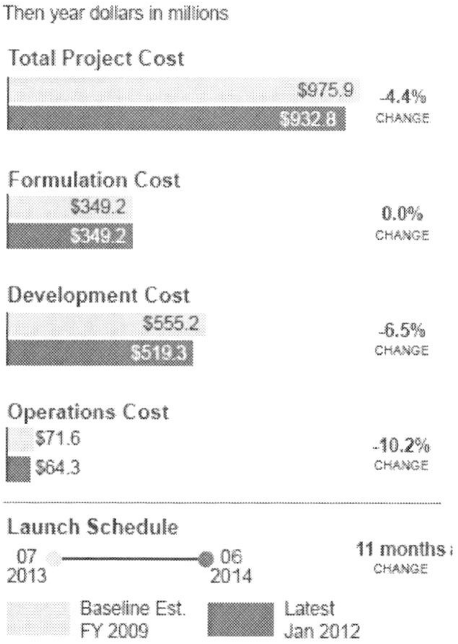

The Global Precipitation Measurement (GPM) mission, a joint NASA and Japan Aerospace Exploration Agency (JAXA) project, seeks to improve the scientific understanding of the global water cycle and the accuracy of precipitation forecasts. The GPM is composed of a core spacecraft carrying two main instruments: a Dual-frequency Precipitation Radar (DPR) and a GPM Microwave Imager (GMI). GPM builds on the work of the Tropical Rainfall Measuring Mission, and will provide an opportunity to calibrate measurements of global precipitation.

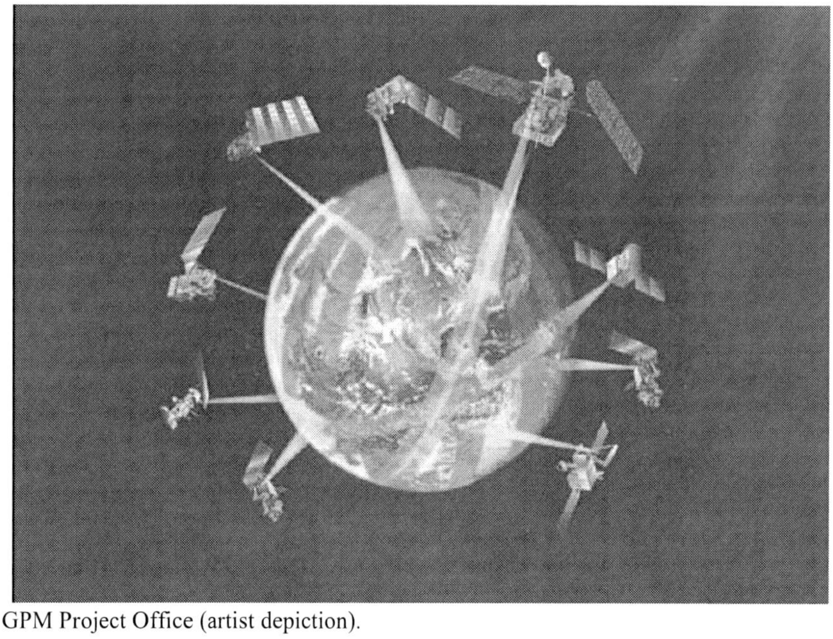

Source: GPM Project Office (artist depiction).

Project Summary

Faced with low schedule reserve for GPM, NASA reported in October 2011 that the project has delayed the launch date 11 months to June 2014. After the March 2011 earthquake in Japan, the project's international development partner experienced component and delivery issues with an instrument that contributed to the delayed launch of the spacecraft. The project is also experiencing late delivery of several components for the spacecraft and the GPM Microwave Imager (GMI-1) instrument, which will result in a delay in the integration and test schedule. The project, in one instance, had to use a second manufacturer for a spacecraft component to mitigate the original contractor's low production capacity.

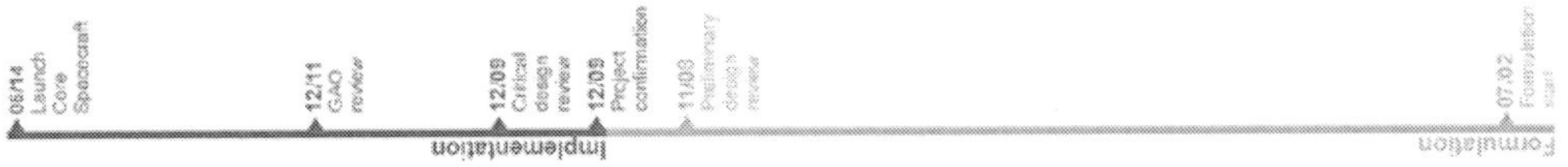

Project Update

Funding Issues

In October 2011, NASA reported that the project had delayed its launch date 11 months from July 2013 to June 2014. In July 2011, the project reported that it was working on an internal replan for a November 2013 launch date. At that time, the project manager stated that the project did not have enough schedule reserve without working weekends to meet the July 2013 launch date, and that this delay primarily stemmed from late delivery of the project's two primary instruments and spacecraft components from its contractors and Japanese development partner. According to NASA, the further delay to June 2014 was due to the earthquake in Japan that further delayed development of the partner-provided instrument as well as additional delays in the spacecraft and GPM Microwave Imager (GMI) development. NASA was also developing a second instrument—GMI-2—where NASA reported investing part of its $32 million in funds received under the American Recovery and Reinvestment Act of 2009 in its development. The President's fiscal year 2012 budget request, however, recommended discontinuing GMI-2 funding. Although the science requirements for GPM can still be met without flying the GMI-2 instrument, project officials reported that without the instrument the available science data from the mission would not be as robust as originally intended.

Development Partner Issues

Project officials stated that NASA has had positive coordination and communication with the Japan Aerospace and Exploration Agency (JAXA), which is providing the Dual-frequency Precipitation Radar (DPR) and the launch vehicle for the GPM spacecraft. After the March 2011 earthquake in Japan, JAXA experienced component and delivery issues with the DPR instrument, which contributed to the second launch delay until June 2014 from launch date of November 2013 in NASA's internal replan. Project officials said that without the DPR they cannot start environmental testing. In June 2011, GPM's program executive reported that the NASA administrator and the JAXA president had agreed to include a secondary JAXA payload on GPM, which could further affect the project's cost and launch date. Project officials stated that JAXA agreed to all GPM requirements and requests for safety reviews and to delete the secondary payload satellites if they present a hazard to the GPM observatory.

Contractor Issues

Prior to the events in Japan, GPM's project schedule had already been affected by the late delivery of the GMI instrument being built by Ball Aerospace and the Power System Electronics and Command and Data Handling unit supplied by Northrop Grumman. For example, the GMI instrument will not be delivered until at least January 2012, which is after it was scheduled to be available for integration and test, due to issues with its radio frequency receivers. Further, given a lack of capacity on the part of Northrop Grumman to produce the Power System Electronics and Command and Data Handling units, according to the project it had to utilize a second manufacturer to help mitigate the late deliveries of these components. The impact of these delays was lessened when the project slipped the launch date to June 2014.

Other Issues to be Monitored

GPM officials reported that a scheduling conflict with the James Webb Space Telescope for test facilities has created risk to the GPM schedule for thermal vacuum testing. The project is also encountering scheduling conflicts with other test facilities for system-level testing and has looked at outside facilities, including non-NASA centers, to mitigate the problem.

PROJECT OFFICE COMMENTS

The GPM project provided technical comments to a draft of this assessment, which were incorporated as appropriate.

Gravity Recovery and Interior Laboratory

Recent/Continuing Project Challenges

- Contractor Issues

Previously Reported Challenges

- Technology Issues
- Launch Issues

Project Essentials

NASA Center Lead: **Jet Propulsion Laboratory (JPL)**
International Partner: **None**
Launch Date: **September 8, 2011**
Launch Location: **Kennedy Space Center, FL**
Launch Vehicle: **Delta II Heavy**

Mission Duration: **9 months**
Requirement derived from:
NASA Strategic Plan

Contract Information

Major Contractor: **Lockheed Martin Corporation**
Type of Contract: **Cost Plus Incentive Fee**
Date of Award: **February 2009**
Initial Value of Contract: **$113.4 million**
Current Value: **$129.3 million**

Project Performance

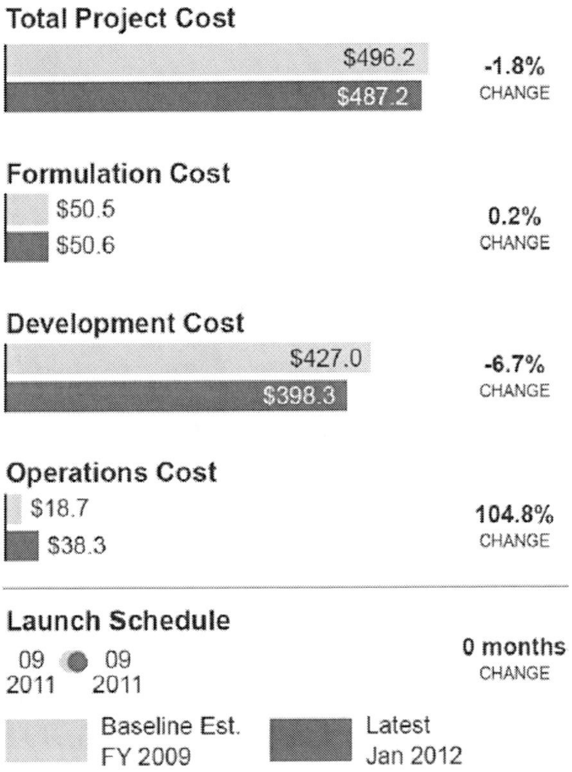

The Gravity Recovery and Interior Laboratory (GRAIL) mission will seek to determine the structure of the lunar interior from crust to core, advance our understanding of the thermal evolution of the Moon, and extend our knowledge gained from the Moon to other terrestrial-type planets. GRAIL will achieve its science objectives by placing twin spacecraft in a low altitude and nearly circular polar orbit. The two spacecraft will perform high-precision measurements between them. Analysis of changes in the spacecraft-to-spacecraft data caused by gravitational differences will provide direct and precise measurements of lunar gravity. GRAIL will ultimately provide a global, high-accuracy, high-resolution gravity map of the Moon.

Source: Courtesy of NASA/JPL-Caltech (artist depiction).

Project Summary

GRAIL successfully launched on September 10, 2011, both on schedule and within cost. In January 2012, NASA reported that the GRAIL project underran its development costs by more than $28 million. The project will use a portion of that savings to accommodate operations costs and provide funding for unknown technical risks. As a result of late avionics deliveries, the project used cost and schedule reserves to offset having to maintain an increased contractor workforce for the entire year prior to launch. The project continues to monitor and negotiate for access to Deep Space Network coverage as the network is expected to be oversubscribed for at least some portion of the project's mission duration.

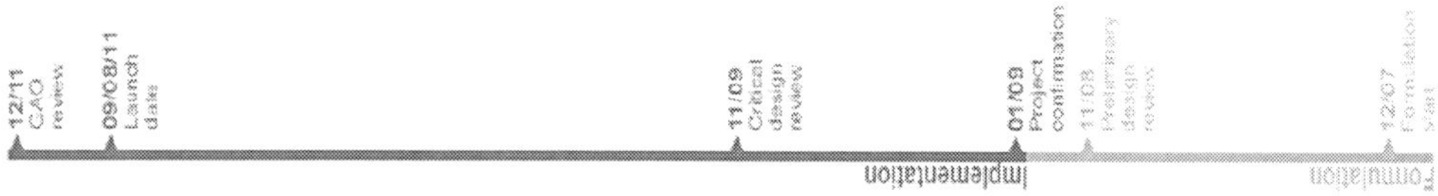

Project Update

GRAIL successfully launched on September 10, 2011, and the twin spacecraft—recently named Ebb and Flow—achieved lunar orbit on December 31, 2011, and January 1, 2012. The project met its commitments within cost and on schedule. In January 2012, NASA reported that the GRAIL project underran its development costs by more than $28 million. The project will use a portion of that savings to accommodate operations costs and provide funding for unknown technical risks.

Contractor Issues

Contractor workforce levels were higher than planned for the year prior to launch. Project officials stated that this was due primarily to late avionics deliveries, including staff to support these late deliveries, and a delay of the software acceptance test program. The project used available cost and schedule reserves to offset the impact of these issues, which the project estimated to be 6 weeks and $11 million. The project used engineering units for test purposes while waiting for the flight units. Project officials stated that using engineering units provided a significant mitigation to additional schedule loss that may have caused cost increases.

PROJECT OFFICE COMMENTS

The GRAIL project provided technical comments to a draft of this assessment, which were incorporated as appropriate. Project officials also commented that GRAIL successfully launched two spacecraft on September 10, 2011, on schedule, within cost, and that the spacecraft were performing well. They added that GRAIL had successfully obtained, or negotiated support for, all required DSN coverage through lunar orbit insertion. The project remains concerned about portions of the transition to science formation; however, project officials expect any scheduling conflicts will be resolved in early 2012.

Other Issues to be Monitored

The project continues to monitor availability of Deep Space Network (DSN) communications resources. According to project officials, the project may have to negotiate with other projects during its late cruise and late transition to science formation periods in order to obtain necessary DSN coverage. They stated that GRAIL project analysis indicated that the DSN will be oversubscribed in the late 2011 and early 2012 time frame. Officials noted that for the science phase of the mission in early 2012, GRAIL revised its tracking coverage requirement from 24 to 16 hours per day at no impact to science. The project has also approved the purchase of additional receivers at the three DSN complexes to receive the tracking signals from both spacecraft simultaneously while in the science phase.

Ice, Cloud, and Land Elevation Satellite-2

Recent/Continuing Project Challenges

- Launch Issues
- Funding Issues

Project Essentials

NASA Center Lead: **Goddard Space Flight Center**
International Partner: **None**
Projected Launch Date: **April - November 2016**
Launch Location: **Vandenberg AFB, CA**
Launch Vehicle: **TBD**
Mission Duration: **3 years (5 year goal)**
Requirement derived from: **2007 Earth Science Decadal Survey**

Contract Information

Major Contractor: **Orbital Sciences Corp.**
Type of Contract: **Fixed Price**

Date of Award: **September 2011**
Initial Value of Contract: **$135.1 million**
Current Value: **$135.1 million**

Project Performance

Then year dollars in millions

* This estimate is preliminary, as the project is in formulation and there is uncertainty regarding the costs associated with the design options being explored. NASA uses these estimates for planning purposes.

Preliminary estimate of Project Life Cycle Cost*

Launch Schedule **04/2016 – 11/2016**

NASA's Ice, Cloud, and Land Elevation Satellite-2 (ICESat-2) is a first-tier mission recommended by the National Research Council in its 2007 Earth Science Decadal Survey. ICESat-2 is a follow-on mission to ICESat, tasked with measuring changes in polar ice-sheet mass with space-borne altimetry measurements to understand mechanisms that drive change and the impact of these changes on future global sea level. ICESat-2 will utilize a micro-pulse multi-beam laser instrument with a photon counting approach to measurement. This process will allow for dense cross-track sampling with a high repetition rate, allowing ICESat-2 to provide better elevation estimates over high slope and rough areas.

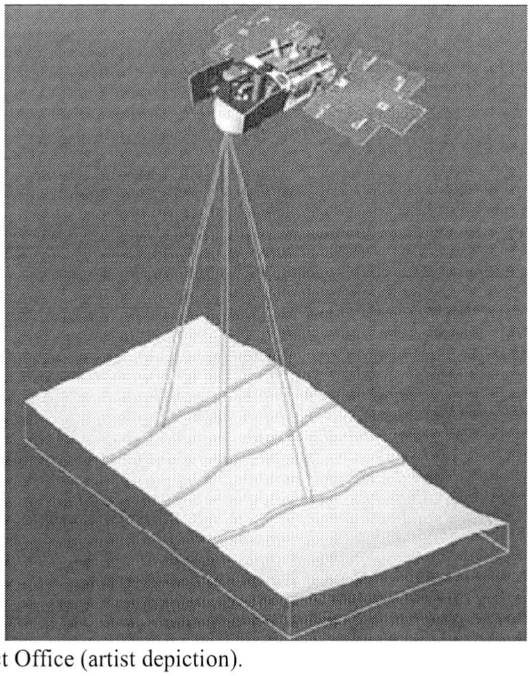

Source: ICESat-2 Project Office (artist depiction).

Project Summary

The launch vehicle for ICESat-2 is currently undefined. NASA is reviewing a plan to launch ICESat-2 with an Air Force mission to allow the project to share launch expenses, potentially saving money and helping the project stay within cost parameters. However, schedule delays with the Air Force mission may result in ICESat-2 procuring its own launch vehicle, which project officials state will add cost to the mission. Although the project has not been confirmed with a baseline cost and schedule, project officials are now reporting an April 2016 launch date, which is 6 months after the preliminary date previously reported.

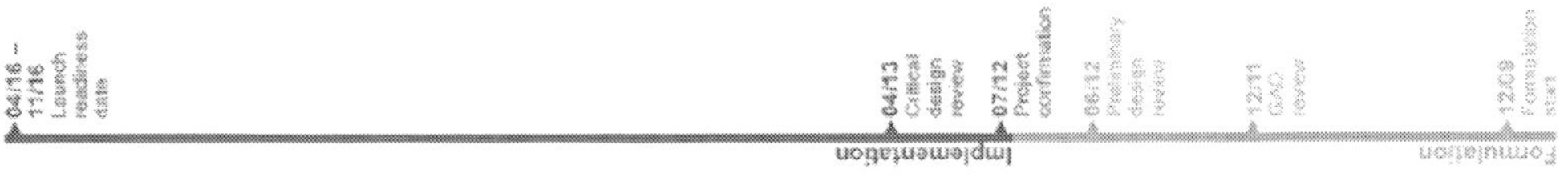

Project Update

In July 2011, ICESat-2 successfully completed a management review and the project was approved to enter Phase B, preliminary design and technology completion. Although ICESat-2 will not have an approved baseline schedule until it is confirmed in July 2012, its planned launch date is 6 months later than previously reported and is now tentatively scheduled for April 2016. The project has begun to award contracts for key project technologies and development. For example, in May 2011 the project awarded a contract to one of the four vendors working to mature micro-pulse laser technology—Fibertek, Inc. Also, in August 2011, the project awarded the contract for spacecraft development to Orbital Sciences Corporation.

Launch Issues/Funding Issues

According to project officials the launch vehicle for ICESat-2 remains undefined because of the high cost of appropriate launch vehicles. In order to reduce the cost of a launch vehicle and stay within its life-cycle cost range estimate, the project is currently pursuing a plan to launch ICESat-2 on the same vehicle—or co-manifest----with an Air Force weather satellite mission. If ICESat-2 were to co-manifest with the Air Force mission, the launch vehicle would be procured by the Air Force. NASA would, however, have a reduced level of oversight and control since the Air Force would be responsible for managing and mitigating launch vehicle risks. There would also be increased schedule and technical challenges with having two separate missions launch at the same time. NASA would be responsible for an estimated $50 million to cover developing a device to enable the two missions to launch together on the same vehicle and project launch site costs. Currently, due to potential budget constraints with Air Force weather satellite missions, it is possible that the Air Force mission would be delayed. If an agreement cannot be reached between the Air Force and NASA for co-manifesting the missions, project officials said they will work with NASA Launch Services to procure a launch vehicle to independently launch ICESat-2, which they added would be significantly more costly.

Project officials stated that, optimally, the project would know the launch vehicle selection by preliminary design review (PDR) because launch vehicle and spacecraft interfaces are developed at that point. They added that the project has flexibility within its spacecraft contract to allow for a delay up to a year in defining the launch vehicle at a fixed price. The project is planning on holding the mission PDR in June 2012, which would allow the project up to June 2013 to select the launch vehicle. NASA officials added, however, that the instrument CDR is planned for October 2012 and any design changes required beyond that point driven by a subsequent launch vehicle selection would drive cost and schedule.

PROJECT OFFICE COMMENTS

The ICESat-2 project provided technical comments to a draft of this assessment, which were incorporated as appropriate. Project officials also commented that the ATLAS instrument completed its preliminary design review in November 2011.

Other Issues to be Monitored

The ICESat-2 project is tracking a risk on the Advanced Topographic Laser Altimeter System (ATLAS) instrument—which is being developed in house by NASA—because of uncertainty in the daily data volume. This issue could lead to the project not allocating enough space for the daily science data to be downloaded and result in a loss of data. To mitigate the potential loss of data, the project is working on a plan to better identify the level of background noise, which can contribute to higher daily data volume, and provide appropriate capacity in the ground system in case the data volume is unexpectedly higher than current predictions.

James Webb Space Telescope

Recent/Continuing Project Challenges

- Funding Issues
- Contractor Issues
- Design Issues
- Technology Issues

Previously Reported Challenges

- Complexity of Heritage Technology

Project Essentials

NASA Center Lead: **Goddard Space Flight Center**

International Partners: **European Space Agency (ESA), Canadian Space Agency (CSA)**

Projected Launch Date: **October 2018**

Launch Location: **Kourou, French Guiana**

Launch Vehicle: **Ariane 5 (ESA Supplied)**

Mission Duration: **5 years (10 year goal)**

Requirement derived from: **Astrophysics Decadal Survey, 2001**

Contract Information

Major Contractor: **Northrop Grumman Aerospace Company**
Type of Contract: **Cost Plus Award Fee**
Date of Award: **2002**
Initial Value of Contract: **$824.8 million**
Current Value: **$1.93 billion**

Project Performance

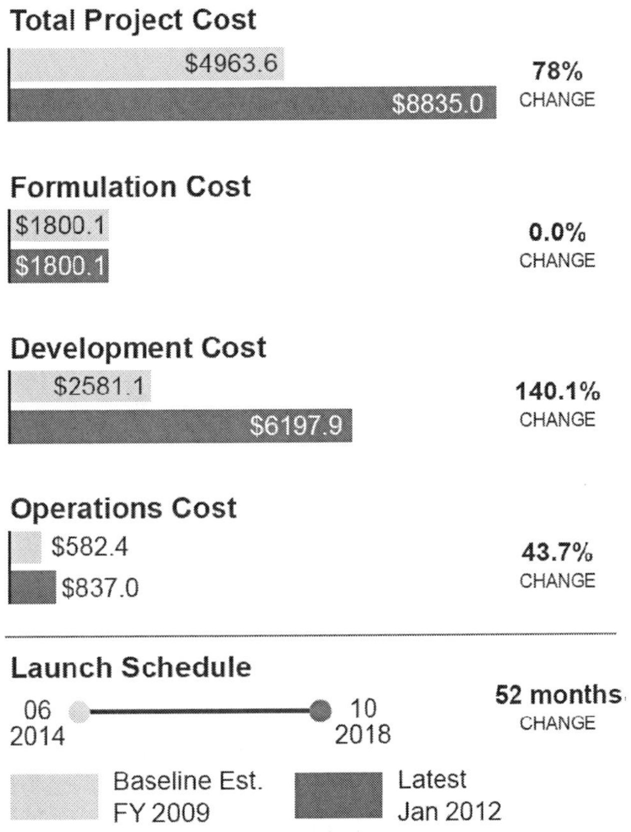

The James Webb Space Telescope (JWST) is a large, infrared-optimized space telescope that is designed to find the first galaxies that formed in the early universe. Its focus will include searching for first light, assembly of galaxies, origins of stars and planetary systems, and origins of the elements necessary for life. JWST's instruments will be designed to work primarily in the infrared range of the electromagnetic spectrum, with some capability in the visible range.

JWST will have a large primary mirror composed of 18 smaller mirrors, measuring 6.5 meters (21.3 feet) in diameter, and a sunshield that is the size of a tennis court. Both the mirror and sunshield will unfold and open once JWST is in outer space. JWST will reside in an orbit about 1.5 million kilometers (1 million miles) from the Earth.

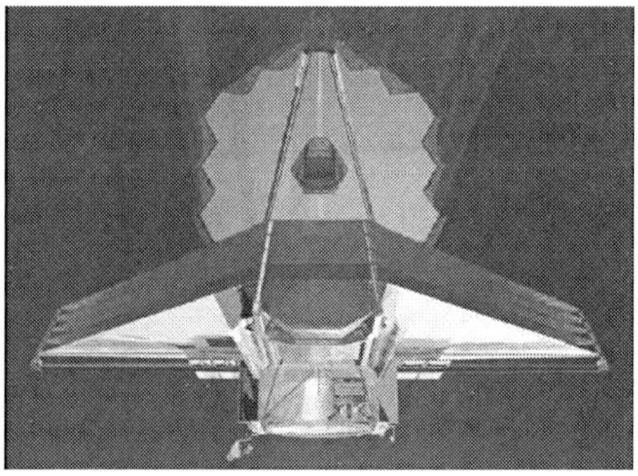

Source: Northrop Grumman Aerospace Systems (artist depiction).

Project Summary

In October 2011, NASA announced that the estimate of the project's total life-cycle cost is now \$8.835 billion with a launch readiness date in October 2018. NASA has restructured JWST's management by taking the lead role in systems engineering from the prime contractor and providing more oversight of contractors developing the instruments. Though much progress has been made, the project has technical challenges remaining with the design of the sunshield and completion of the spacecraft bus, which was delayed as the project worked on higher priority instrument development efforts. The project has experienced instrument delivery and design issues that have delayed integration and testing efforts.

Project Update

Funding Issues

In 2011, the JWST program underwent a replan in response to an Independent Comprehensive Review Panel report issued in October 2010. In October 2011, NASA reported that the estimate of the project's total life-cycle cost is now \$8.835 billion—which includes a development cost increase of 140 percent from its baseline with a launch readiness date of October 2018, a delay of 52 months. NASA and Northrop Grumman officials reported that the project, as part of the replan efforts, was to assume a constrained budget environment for fiscal years 2011 and 2012 and unconstrained budgets after that. In early 2011, project officials called for the contractor and subcontractors to cease reporting earned value management (EVM) data until the replanning effort is complete and the project is rebaselined. NASA officials also reported that there are no descoped options available that would still allow it to meet the science requirements. According to NASA officials, other projects, starting with those that have not yet been confirmed, may face possible cancellation to help offset JWST's costs. The future of the project was debated within Congress as the House Appropriations Committee recommended terminating funding for JWST, while the Senate provisionally funded the program in its bill but proposed a development cost cap of \$8 billion for the project. In November 2011, Congress funded the project for fiscal year 2012 with an \$8 billion cost cap for the formulation and development phases of the project.

Contractor Issues

The JWST project instituted several changes to the work being performed by its contractors. For example, the project has changed the award fee structure for the prime contract with Northrop Grumman to make a greater percentage of the award fee dependent on meeting the cost target, according to NASA. NASA also reported that it has assumed the lead role for the systems engineering functions of the program as well as some of the integration and testing responsibility from Northrop Grumman as it believes it is better suited to perform these tasks, according to NASA. Northrop Grumman, however, will maintain the systems engineering role over the spacecraft elements that it has under contract for development, according to both NASA and Northrop Grumman. Project officials also stated that because of Lockheed Martin's struggles in developing the near infrared camera and changes in the Lockheed Martin contractual requirements, the delivery of the instrument has been delayed and is over budget, resulting in a cost increase between \$60 million and \$200 million. In addition, the project sent two representatives to the contractor facility for increased oversight. The increase in cost for the near infrared camera was encompassed in the replan.

Design /Technology Issues

In September 2011, the project announced that it had completed polishing and coating all of the 18 primary mirrors. Technical challenges, however, remain in developing the sunshield, according to NASA. For example, NASA officials stated that challenges with the sunshield are making sure the membrane does not rip inside the fairing during liftoff and that it unfolds properly in space. Project officials also told us that work on the spacecraft was delayed while the project was working on the higher risk development of the instruments, most of which are now complete. In addition, the project reported problems affecting the instruments that will be included on the Integrated Science Instrument Module (ISIM) would have caused a schedule delay, but were incorporated in the replan. These problems included degradation of near-infrared detectors and structural cracks found in the near infrared spectrograph with potential impact to its functionality. According to the project, although the ISIM structure began integration and testing in the summer of 2011, the delay in the delivery of the spacecraft bus will likely cause the project to store the primary mirrors and the ISIM instruments for up to 43 and 31 months, respectively, until integration and testing with the spacecraft begins. Project officials report that they have no concerns about the long periods of inactivity for these components

PROJECT OFFICE COMMENTS

The JWST project provided technical comments to a draft of this assessment, which were incorporated as appropriate.

Juno

Recent/Continuing Project Challenges

- Design Issues

Previously Reported Challenges

- Technology Issues
- Contractor Issues
- Development Partner Issues
- Parts Issues

Project Essentials

NASA Center Lead: **Jet Propulsion Laboratory (JPL)**

International Partners: **Agencia Spaziale Italiana - Selex Galileo; ASI – Thales Alenia Space, Centre Spatial de Liege Belgian Science Policy, Centre National d'Etudes Spatiales – Centre d'Etude**

Launch Date: **August 5, 2011**

Launch Location: **Cape Canaveral AFS, FL**

Launch Vehicle: **Atlas V**

Mission Duration: **6 years**

Requirement derived from: **New Frontiers Announcement of Opportunity**

Contract Information

Major Contractor: **Lockheed Martin**

Type of Contract: **Cost Plus Award Fee/Incentive Fee**

Date of Award: **August 2008**

Initial Value of Contract: **$195.0 Million**

Current Value: **$251.1 Million**

Project Performance

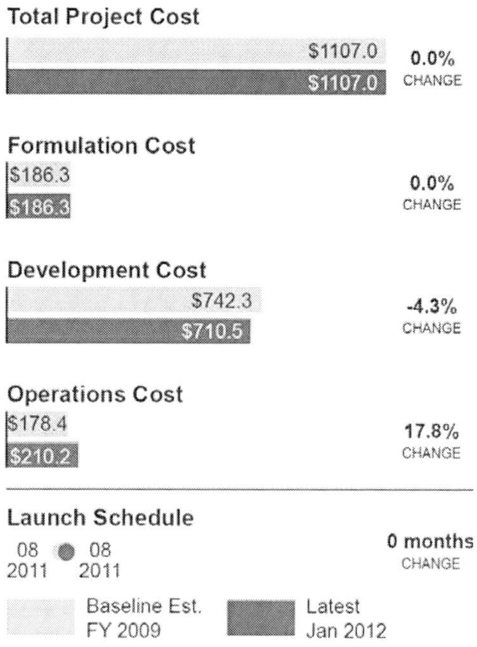

Source: NASA/JPL (artist depiction).

The Juno mission seeks to improve our understanding of the origin and evolution of Jupiter. Juno plans to achieve its scientific objectives by using a simple, solar-powered spacecraft to make global maps of the gravity, magnetic fields, and atmospheric conditions of Jupiter from a unique elliptical orbit. The spacecraft carries precise, highly sensitive radiometers, magnetometers, and gravity science systems. Juno is slated to make 32 orbits to sample Jupiter's full range of latitudes and longitudes. From its polar perspective, Juno is designed to combine local and remote sensing observations to explore the polar magnetosphere and determine what drives Jupiter's auroras.

Project Summary

Juno successfully launched on August 5, 2011, and began its 5-year cruise phase to Jupiter. Although the project launched within cost and on schedule, it exhausted most of the project-level reserves and may request additional funding from headquarters for post-launch development activities. Late delivery of components led the project to use cost and schedule reserves to offset the impact of delayed testing and increased workforce to meet the scheduled launch date. During the initial checkout process of Juno's instruments and sensors, it was determined that two of its components are running at higher than expected temperatures, due in part to a radiator not being placed in the recommended position.

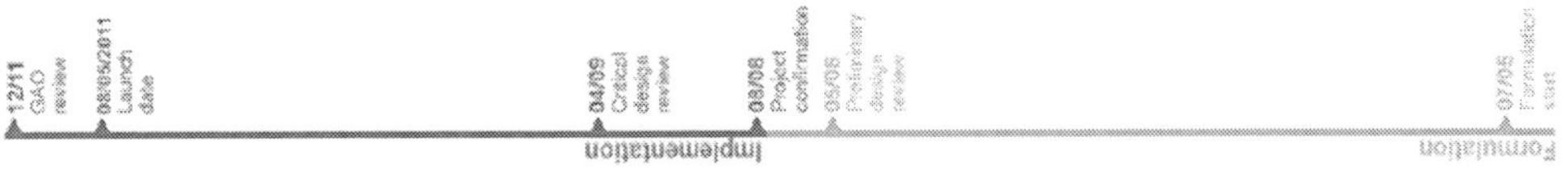

Project Update

The Juno project successfully launched on August 5, 2011, and expects to reach Jupiter in 2016. The project is performing initial status checks on the spacecraft and instruments. The project met its commitments within cost and on schedule.

The project exhausted most of its project-held reserves due, largely, to increased workforce and unanticipated technical challenges. Due to the late delivery of the avionics, environmental testing of Juno was delayed and completed 6 weeks behind schedule. As a result, the initial integration phase of the project took longer than planned and required additional workforce shifts to maintain the schedule margin within acceptable levels. After using most of its project-level reserves, Juno received an additional $15 million of unallocated future expenses in April 2011 from the Science Mission Directorate to cover costs through the launch date. Despite this added funding, the project acknowledges it may need to request additional funds. Project officials reported that the additional funding will allow the project to carry over reserves to the operations phase in order to support post-launch development activities. In January 2012, NASA reported that $18 million in unused reserves from development were being carried forward to the operations phase.

Design Issues

Two of Juno's components—the Jovian Auroral Distributions Experiment–Ion (JADE I) sensor and the Jupiter InfraRed Auroral Mapper (JIRAM) detector—are running at higher temperatures than predicted on-orbit. The project has found that part of the temperature difference for JADE I is due to a discrepancy in the radiator placement. The project's thermal team had recommended that the radiator be placed in one location on the sensor, but it was built in another. According to project officials, a build team worker determined that the radiator would not fit in the recommended position due to overcrowding of components; however, the change in location was not communicated back to the thermal team. The project team is currently updating its thermal model to uncover additional heating sources to attempt to explain the remaining temperature difference. The project has delayed the instrument checkout until the spacecraft has reached a location where the sensor can be turned on at a temperature below its operating limit. The project is concerned whether the JADE I sensor will maintain calibration as it orbits close to the sun during its cruise to Jupiter due to the increase in the thermal environment. If the JADE I sensor does not maintain its calibration, it could result in reduced quality of science data collected. The JIRAM detector is a temperature sensitive component and, while it is not needed to meet the project science requirements, a warmer than expected detector may have an impact on science data quality. The project manager said that the project is currently working to determine the root causes for the JIRAM detector's overheating issues.

The Juno project had released only 39 percent of the engineering drawings at the critical design review (CDR). Project officials, however, said they were able to budget for and use engineering models for all instruments to demonstrate design maturity at CDR. For some spacecraft components, the Juno project did not build or test engineering models because they were heritage designs. For example, some spacecraft components being utilized are very similar to the ones used on the Mars Reconnaissance Orbiter; therefore, the project accepted some of the spacecraft designs based on qualification testing.

Project Office Comments

The Juno project provided technical comments to a draft of this assessment, which were incorporated as appropriate. Project officials also commented that Juno completed its development within cost and on schedule and launched successfully on the first day of its launch period.

Landsat Data Continuity Mission

Recent/Continuing Project Challenges

- Funding Issues
- Development Partner Issues
- Parts Issues

Previously Reported Challenges

- Technology Maturity

Project Essentials

NASA Center Lead: **Goddard Space Flight Center**
Partner: **U.S. Geological Survey (USGS)**
Projected Launch Date: **June 2013**
Launch Location: **Vandenberg AFB, CA**
Launch Vehicle: **Atlas V**
Mission Duration: **5 years (10 years propellant)**
Requirement derived from: **Continuation of Landsat data series, 1972**

Contract Information

Major Contractor: **Ball Aerospace and Technologies Corp.**
Type of Contract: **Cost Plus Award Fee**
Date of Award: **July 2007**
Initial Value of Contract: **$127.8 million**
Current Value: **$188.7 million**

Project Performance

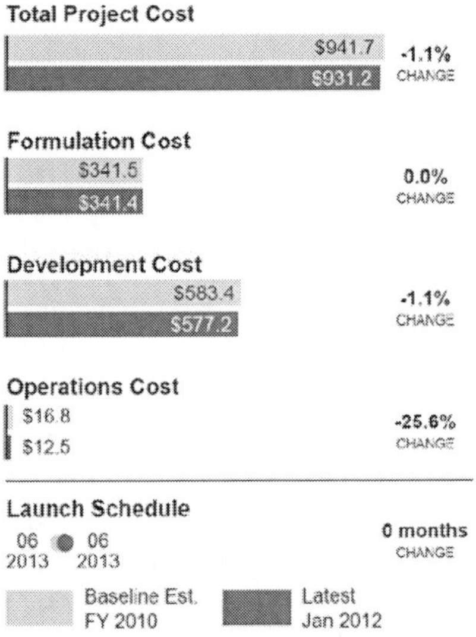

Source: Orbital (artist depiction).

The Landsat Data Continuity Mission (LDCM), a partnership between NASA and the United States Geological Survey (USGS), seeks to extend the ability to detect and quantitatively characterize changes on the global land surface at a scale where natural and man-made causes of change can be detected and differentiated. It is the successor mission to Landsat 7. The Landsat data series, begun in 1972, is the longest continuous record of changes in the Earth's surface as seen from space. Landsat data is a resource for people who work in agriculture, geology, forestry, regional planning, education, mapping, and global change research.

Project Summary

The LDCM project continues to work toward a December 2012 launch in order to avoid or minimize a gap in Landsat data. However, LDCM is facing technical and schedule hurdles that may result in the project launching in December 2012 without fully testing the Thermal Infrared Sensor instrument (TIRS), which would increase project risk of failing to meet mission requirements. The project experienced delays due, in part, to design issues in the launch lock mechanism on TIRS, and delays getting parts for the spacecraft. The project is also concerned about the funding stability of LDCM's ground system partner, the United States Geological Survey (USGS).

Project Update

LDCM is facing technical and schedule hurdles that project officials stated may result in the project launching in December 2012 without fully testing the TIRS instrument, increasing project risk of failing to meet mission requirements. TIRS was added to LDCM during the confirmation process with an estimated additional cost to the project of $160 million. The Standing Review Board acknowledged at that time that the schedule for TIRS was aggressive to meet a December 2012 launch date, and recommended establishing a baseline launch date of June 2013, while continuing to internally work towards December 2012. Project officials stated that launching in December 2012 is essential because of the risk of a gap in science data if the Landsat 5 and 7, which are both currently operating well past their design lives, cease operations.

Funding Issues

LDCM is less than a year from its internal launch date, and project officials are using reserve funding held at headquarters to finish TIRS development. TIRS development consumed the cost reserves held by the project for fiscal year 2011 and resulted in the need for the project to receive cost reserves held at NASA headquarters. Project officials stated that TIRS has used an average of approximately $1.5 million of project-held reserves per month during the last year. The TIRS instrument is going to be delivered late, in part, because of an issue with the launch lock releasing correctly and unexpected performance of the instrument during thermal vacuum chamber testing. Despite the use of reserve funding, NASA recently reported that it was lowering the LDCM lifecycle cost estimate by 1.1 percent due to retirement of key cost risks pertaining to the delivery and integration of the Operational Land Imager instrument and completion of TIRS environmental testing. Project officials stated that the total Mission Directorate-managed unallocated future expenses are enough funding to cover project costs in the event of a launch delay into 2013, but potential launch services costs for such a delay could exceed this amount. Project officials stated, for example, that potential launch services costs associated with the change in launch date might cause the project to exceed its total life cycle cost. Project officials also stated concern that the launch manifest in 2013 is crowded and that the project could be further delayed past the June 2013 baseline date as a result.

Development Partner Issues

The project stated concern about the funding stability of LDCM's ground system partner, the USGS. As we previously reported, USGS had amended its agreement with NASA to defer some of USGS's development funding responsibilities to operational funding in later years. NASA officials stated that it is unclear whether USGS will have the budget to fully fund its commitment for the remaining development phase and operations phase after launch.

PROJECT OFFICE COMMENTS

The LDCM project provided technical comments to a draft of this assessment, which were incorporated as appropriate. Project officials also commented that all threshold (minimum) requirements can be achieved solely by the Operational Land Imager. They added that they expect TIRS will be fully tested, satisfy its baseline requirements, and delivered in sufficient time to meet the current launch readiness date. Project officials also reported that, as a schedule driven mission committed to an internal launch readiness date of December 2012, LDCM requested additional reserves in December 2010 as a proactive approach to mission management across the project to meet this schedule.

Parts Issues

The project also experienced delays getting flight components for the spacecraft, including the star tracker and the flight payload interface electronics, necessitating workarounds in testing, with potential risks to the observatory integration and testing timeline. The star tracker delay was, in part, a result of instability exhibited during testing. The star tracker also had an issue with chips on a data processing board that had to be replaced; this issue was reported by other projects as well.

Lunar Atmosphere and Dust Environment Explorer Formulation

Recent/Continuing Project Challenges

- Technology Issues
- Launch Issues
- Design Issues

Previously Reported Challenges

- Parts Issues

Project Essentials

NASA Center Lead: **Ames Research Center**
International Partners: **None**
Projected Launch Date: **November 2013**
Launch Location: **Wallops Flight Facility, VA**
Launch Vehicle: **Minotaur V**
Mission Duration: **180 days**
Requirement derived from: **National Research Council**

Contract Information

Major Contractor: **In-house development**
Type of Contract: **N/A**
Date of Award: **N/A**
Initial Value of Contract: **N/A**
Current Value: **N/A**

The Lunar Atmosphere and Dust Environment Explorer (LADEE) mission objective is to determine the global density, composition, and time variability of the lunar atmosphere. LADEE's measurements will determine the size, charge, and spatial distribution of electrostatically transported dust grains. Additionally, LADEE will carry an optical laser communications demonstrator that will test high-bandwidth communication from lunar orbit.

Source: LADEE Project Office (artist depiction).

Project Performance

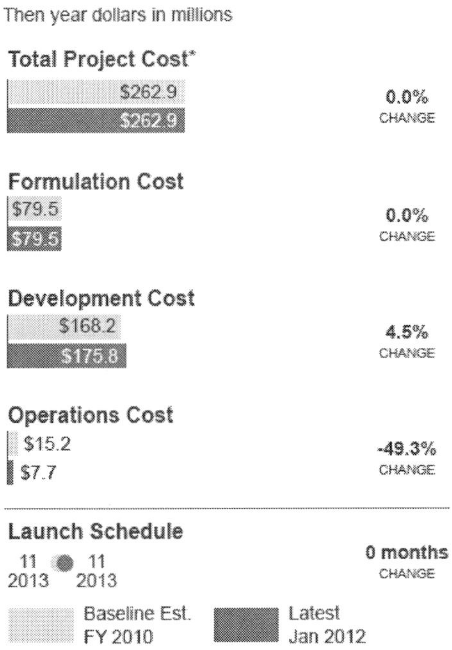

* This estimate does not include the LLCD instrument which is being funded by the Human Exploration and Operations Mission Directorate at a cost of approximately $65 million.

Project Summary

The LADEE project did not pass its recent system integration review and a second review will be required for the project to move forward. The LADEE project is working toward an internal launch date of May 2013, six months earlier than its baseline launch date. The May 2013 launch is at risk because development of the Lunar Laser Communications Demonstration has fallen behind. The LADEE project has also experienced delays due to a bid protest of the Minotaur V launch vehicle selection in 2009. As a result, analysis of the launch environment was also delayed. When that analysis was performed, it showed the need for modifications to the spacecraft to withstand the loads produced by the launch vehicle.

Project Update

The project did not pass its system integration review (SIR) in November 2011 largely, according to a senior NASA official, because of a lack of experience developing test and integration procedures within the project and at Ames Research Center and that the review was held earlier than is typical for other projects. The official stated that the impact of the SIR failure on the project's cost and schedule are not clear at this time; however, additional manpower will be needed to address the test and integration issues, and a second systems integration review in early 2012 is required for the project to move forward.

Technology Issues

NASA will fly the Lunar Laser Communications Demonstration (LLCD) as a ride along technology demonstration on the LADEE mission. LLCD is being developed by the Human Exploration and Operations Mission Directorate at a cost of approximately $65 million—an additional cost not included in the LADEE life-cycle cost estimate. The project maintains that LADEE will launch with or without the LLCD, but the agency does have a strong desire to launch with the LLCD. Development of the LLCD has fallen behind, threatening the project's internal management launch date in May 2013. As a result, a partner agency has agreed to pay for additional work on the LLCD, including providing funding for the development of an engineering model to begin testing. The project reported that the engineering model is being used for early integration testing to reduce risk. The project anticipates the LLCD will complete integration on schedule.

Launch Issues

LADEE will be the first NASA mission to be launched on a Minotaur V, which was procured under an Air Force contract. NASA has a reduced level of oversight and control with the Minotaur V because it does not have to be certified for use by NASA projects since the Air Force is responsible for providing launch service mission success oversight for LADEE. Furthermore, though GAO denied the protest, the filing of a bid protest regarding the selection of the Minotaur V delayed the project's coupled loads analysis to determine the launch environment of the Minotaur V that LADEE's design will need to accommodate. Project officials stated that, generally, a project can utilize the coupled loads analysis data from previous project launches. Because LADEE is the first mission to launch on a Minotaur V, there was no previous loads analysis data available to facilitate design efforts. Project officials told us that when they received the loads analysis data and ran a new analysis for LADEE, it showed insufficient strength margins and the project had to redesign portions of the spacecraft. Structural manufacturing of the spacecraft to meet the coupled loads data was a concern and project officials said this effort used a significant amount of fiscal year 2011 reserves to complete. The project reports that the flight structure design and manufacturing has been completed and the flight structure was delivered to Ames in December 2011.

Design Issues

The LADEE project continues to monitor the mass of its instruments and the spacecraft's components to ensure that it remains within the bounds of the launch vehicle margins. The project was below the Ames Research Center requirement of 15 percent approaching the critical design review, but has since remained above the required mass margin. Mass is a particular concern for the LADEE project because the mission will be launched at its maximum performance and maximum fuel levels, unlike other missions that may not need this level of performance and are able to trade fuel mass for spacecraft mass.

Other Issues to be Monitored

According to project officials, they are concerned with the number of quality assurance staff available at Ames Research Center, since LADEE is the first in-house space flight development project at that center. According to NASA officials, they have taken steps to help mitigate this issue by sending staff from Goddard Space Flight Center, Glenn Research Center, and Kennedy Space Center to supplement the staff at Ames.

PROJECT OFFICE COMMENTS

The LADEE project provided technical comments to a draft of this assessment, which were incorporated as appropriate.

Magnetospheric Multiscale

Recent/Continuing Project Challenges

- Funding Issues
- Contractor Issues
- Design Issues
- Part Issues

Previously Reported Challenges

- Development Partner Issues
- Technology Issues

Project Essentials

NASA Center Lead: **Goddard Space Flight Center**
International Partners: **Austria, France, Japan, Sweden**
Projected Launch Date: **March 2015**
Launch Location: **Cape Canaveral AFS, FL**
Launch Vehicle: **Atlas V**
Mission Duration: **2 years**
Requirement derived from: **Solar and Space Physics Decadal Survey, 2003**

Contract Information

Major Contractor: **Southwest Research Institute**
Type of Contract: **Cost Plus Fixed Fee**
Date of Award: **April 2003**
Initial Value of Contract: **$225 million**
Current Value: **$225 million**

The Magnetospheric Multiscale (MMS) is made up of four identically instrumented spacecraft. The mission will use the Earth's magnetosphere as a laboratory to study the microphysics of magnetic reconnection, energetic particle acceleration, and turbulence. Magnetic reconnection is the primary process by which energy is transferred from solar wind to Earth's magnetosphere and is the physical process determining the size of a space weather storm. The four spacecraft will fly in a pyramid formation, adjustable over a range of 10 to 400 kilometers. The data from MMS will be used as a basis for predictive models of space weather in support of exploration.

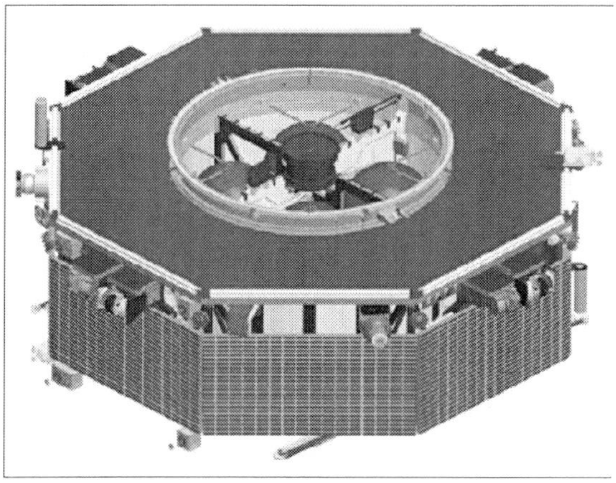

Source: MMS Project Office (Computer Model).

Project Performance

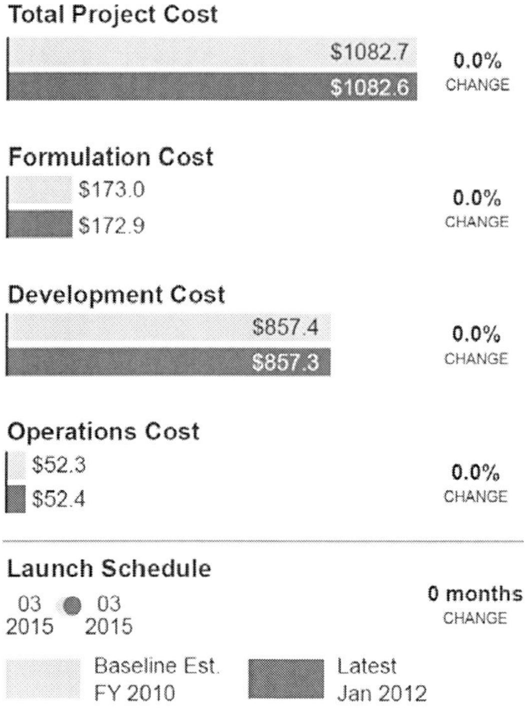

Project Summary

The MMS project is working toward an internal management launch date of August 2014, which would allow the mission to improve its opportunities to conduct science. To meet this earlier launch date, the project used much of its project-held cost reserves and it plans to receive an additional $35 million from NASA headquarters for fiscal years 2011 through 2013. Currently, the project is tracking risks related to the Fast Plasma Investigation (FPI) instrument. Parts issues have depleted reserves held for development of the FPI and forced the project to modify its system level integration and test schedule. The project is also reported problems with the design of a sensor that may not survive the launch environment, potentially leading to the loss of data.

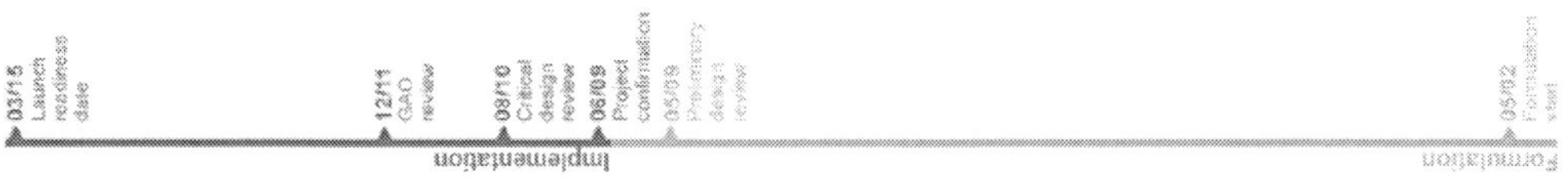

Project Update

Funding Issues

Project officials stated that the MMS project is working toward an internal management launch readiness date of August 2014, which is 7 months earlier than the project's baseline launch date of March 2015. Project officials reported, however, that the launch manifest is currently oversubscribed in the 2014 time frame. Project officials said that the earlier date would allow the mission a window to enter the magnetosphere and improve the project's opportunities to conduct scientific measurements. The project also reported that late deliveries of some instruments have caused cost overruns, and as a result, the project does not have the budget to delay the launch to March 2015 without additional funding from the Science Mission Directorate. To address technical and production problems, the project used much of its project-held cost reserves and plans to receive an additional $35 million from NASA headquarters for fiscal years 2011 through 2013. As of October 2011, the project was holding a reserve posture of approximately 16 percent of the cost remaining, well below the 25 percent required by Goddard Space Flight Center standards, but which project officials consider adequate. Project officials stated that they have options to help mitigate any additional cost overruns. For example, project officials said they could accept more project risk by reducing testing, removing some of the instruments on each observatory, or removing one of the four spacecraft.

Contractor Issues

The MMS project is tracking risks related to the development of the Fast Plasma Investigation (FPI) instrument. As of October 2011, the project reported that the FPI instrument contractor had no remaining reserves for this instrument. Recent detailed cost estimates indicate that the cost to complete its development will exceed the available budget by $15.9 million. According to project officials, the FPI contractor was slow to bring on mechanical designers because it could not find qualified personnel, contributing to delays in developing mechanical drawings and delivering hardware. The project reported that schedule slips in the FPI development have forced the project to modify its system level integration and test schedule to accept a later delivery of this instrument. For example, project officials said that they can now accept the FPI up to a year later than originally planned and meet the August 2014 launch date due to modifying the schedule.

Design Issues

MMS project officials reported problems with the sensor in the accelerometer, a device to measure acceleration, which may not survive the launch environment or the shock when the spacecraft separates from the launch vehicle. In addition, the project reported that shifts in the alignment of the sensor during launch or separation may result in a loss of precision and efficiency in orbital maneuvers. Shock testing has indicated that isolation of the accelerometer to absorb the shock is required and the vendor recently made changes to the engineering model to accommodate isolators.

Parts Issues

The MMS project has experienced various electrical, electronic, and electromechanical (EEE) parts issues and electronic circuit board fabrication issues that have already required more than $600,000 to address. Most of this amount was used on an issue with the optocoupler, used in high voltage power supplies on a couple of the instruments, that project officials said delayed delivery of the FPI to the instrument suite by about 2 months at a cost of about $500,000. The octocoupler problem was resolved by a combination of design and process manufacturing changes made by the parts vendor, such as using a different substance with better thermal properties in the octocoupler.

PROJECT OFFICE COMMENTS

The MMS project provided technical comments to a draft of this assessment, which were incorporated as appropriate. Project officials also commented that the problems with EEE parts and external manufacturing supply chain for electronic circuit boards have been surmounted and the project is now engaged in the production of the multiple copies of flight instruments and spacecraft subassemblies required for the mission.

Mars Atmosphere and Volatile EvolutioN

Recent/Continuing Project Challenges

- Design Issues
- Parts Issues

Previously Reported Challenges

- Launch Issues

Project Essentials

NASA Center Lead: **Goddard Space Flight Center**

International Partner: **Institute of Research for Astrophysics and Planetology, Toulouse, France**

Projected Launch Date: **November 18, 2013**

Launch Location: **Cape Canaveral AFS, FL**

Launch Vehicle: **Atlas V**

Mission Duration: **10 months of travel, 1 year of operations**

Requirement derived from: **Competitively selected in 2008 under the Mars Scout 2006 Announcement of Opportunity**

Contract Information

Major Contractor: **Lockheed Martin**
Type of Contract: **Cost Plus Award Fee**
Date of Award: **April 2009**
Initial Value of Contract: **$237 million**
Current Value: **$247 million**

Project Performance

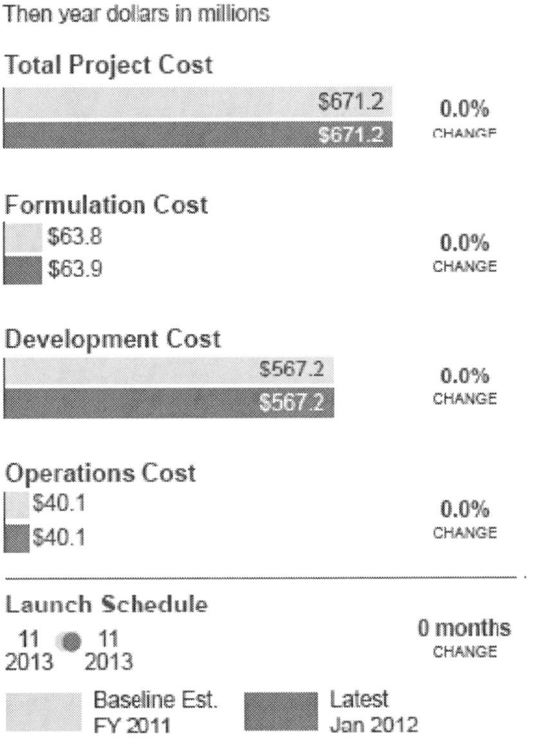

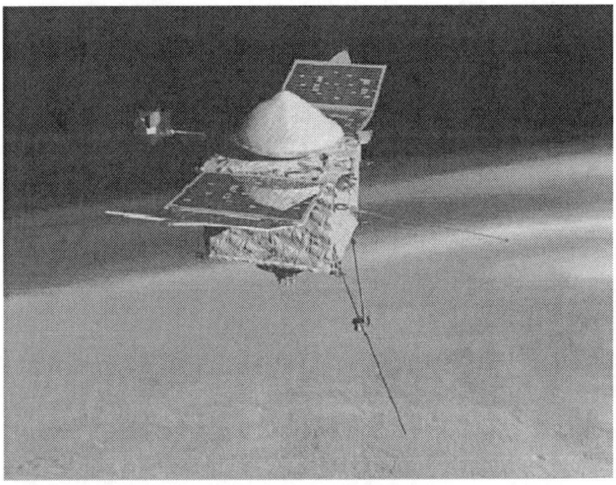

Source: NASA GSFC MAVEN Project Office (artist depiction).

The Mars Atmosphere and Volatile EvolutioN (MAVEN) mission, a robotic orbiter mission, will provide a comprehensive picture of the Mars upper atmosphere, ionosphere, solar energetic drivers, and atmospheric losses. MAVEN will deliver comprehensive answers to long-standing questions regarding the loss of Mars' atmosphere, climate history, liquid water, and habitability. MAVEN will provide the first direct measurements ever taken to address key scientific questions about Mars' evolution.

Project Summary

MAVEN is currently on target to meet its cost and schedule commitments with adequate cost and schedule reserves. The project held its CDR in July 2011 and the Standing Review Board rated the project successful in seven of nine categories. The remaining two categories—design and schedule—were assessed as mostly successful. The review board rated the design drawings as mostly successful because the project did not meet its recommended 80 percent of design drawings complete at CDR. Project officials stated that MAVEN is behind schedule in software development due to delays transitioning staff from two NASA projects that launched in late summer 2011.

Project Update

The MAVEN project is currently on target to meet its cost and schedule commitments with adequate cost and schedule reserves. The project is currently carrying a higher level of cost and schedule reserves for a project in Phase C than is normally required by Goddard Space Flight Center standards. Meeting the project's launch window that begins in November 2013 is critical because MAVEN is a planetary mission and would incur a 26-month delay if the November/December 2013 launch window is missed. Project officials attributed the project's success thus far to factors including having the right people on the project, making tough decisions to prevent requirements creep, having the proper level of funding reserves, and the use of heritage technology.

Design Issues

The MAVEN project held its mission critical design review (CDR) as scheduled in July 2011. The project was rated successful on seven out of the nine categories reviewed by the Standing Review Board, and mostly successful on two other categories—design and schedule. The board rated MAVEN separately on design maturity indicators, and rated the design drawings as mostly successful because the project did not meet their recommended 80 percent of design drawings complete at CDR. We calculated that only 49 percent of the project's current expected engineering drawings were releasable at CDR, which is below GAO's best practices metric of having 90 percent of engineering drawings releasable at CDR.

Two of the top project-reported design risks are the High Efficiency Power Supply (HEPS) Card, and the Neutral Gas and Ion Mass Spectrometer (NGIMS). Last year, project officials were concerned with the high probability of failure of the HEPS, which is MAVEN's power supply system, particularly because it represents a single-point failure. Rather than rework the card, the project obtained a spare HEPS card from NASA's Juno project to conduct early evaluation, and will additionally conduct inspections and testing. This risk will remain open as the contractor tests and inspects the new card. The project also completed a plan to calibrate the NGIMS during on-orbit operations to address a risk that winds in Mars' atmosphere could degrade the science data collected by the instrument. Another risk that project officials stated has been trending downward is the spacecraft launch mass growth. Project officials stated that this risk has been downgraded as the current mass margin has been holding constant, slightly above the 15 percent margin requirement.

Parts Issues

Project Officials reported that an increase in the price for electrical, electronic, and electromechanical parts has produced increases industry-wide, and that parts cost increases are a significant element of the $840,000 projected growth in the cost of the Langmuir Probes and Waves instrument. In addition, the project has identified that the spacecraft propulsion system may affect integration and test because some parts for the system, including the filters, have long lead times from order to delivery. According to project officials, however, the project currently has adequately funded schedule margin and slack to account for this issue.

Other Issues to be Monitored

Project officials stated that MAVEN is behind schedule in flight software development due to delays in transitioning contractor staff from the Juno and Gravity Recovery and Interior Laboratory projects that launched in August and September of 2011, respectively. Project officials stated that they have ramped up the software development staff above the planned levels to make up for previous understaffing and the development is tracking to the recovery plan with no schedule delay.

Project Office Comments

The MAVEN project office provided technical comments to a draft of this assessment, which were incorporated as appropriate. Project officials also commented that the project is making steady progress on releasing drawings, the HEPS and NGIMS design concerns, and the software development with personnel from recently launched missions. They added that the project is on target to conduct the System Integration Review next summer.

Mars Science Laboratory

Recent/Continuing Project Challenges

- Design Issues

Previously Reported Challenges

- Technology Maturity

- Complexity of Heritage Technology
- Parts Issues

Project Essentials

NASA Center Lead: **Jet Propulsion Lab**

Partners: **U.S. Department of Energy, Centre Nationale d'Etude Spatiale (France), Russian Federal Space Agency, Centro de Astrobiologia (Spain), Canadian Space Agency**

Launch Date: **November 26, 2011**

Launch Location: **Cape Canaveral AFS, FL**

Launch Vehicle: **Atlas V**

Mission Duration: **1 year of travel, 2 years of operations**

Requirement derived from: **Part of the Mars Exploration program in response to NASA's Strategic Goals**

Contract Information

Major Contractor: **In-house development**

Type of Contract: **N/A**

Date of Award: **N/A**

Initial Value of Contract: **N/A**

Current Value: **N/A**

Project Performance

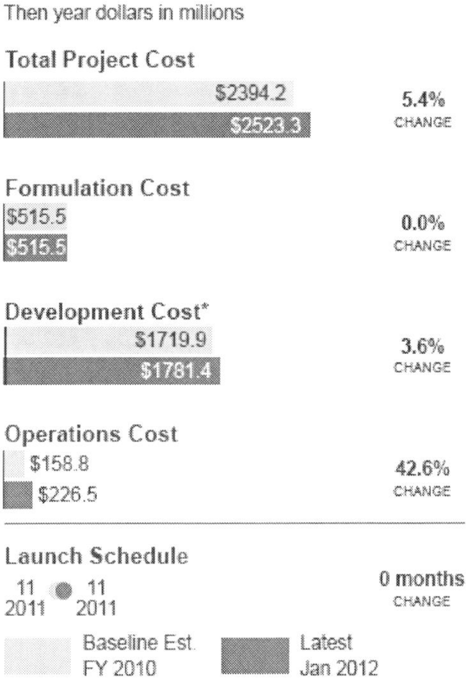

* Represents an 84% growth in development costs since the original baseline of $968.6 established in fiscal year 2008.

Source: NASA/JPL-Caltech.

The Mars Science Laboratory (MSL)—also known by the name of its rover, Curiosity—is part of the Mars Exploration Program (MEP), which seeks to understand whether Mars was, is, or can be a habitable world. To answer this question the MSL project will investigate how geologic, climatic, and other processes have worked to shape Mars and its environment over time, as well as how they interact today. The MSL will continue this systematic exploration by placing a mobile science laboratory on the Mars surface to assess a local site as a potential habitat for life, past or present. The MSL is considered one of NASA's flagship projects and will be the most advanced rover yet sent to explore the surface of Mars. Curiosity is about twice as long and five times as heavy as NASA's twin Mars Exploration Rovers, Spirit and Opportunity, launched in 2003.

Project Summary

The Mars Science Laboratory successfully launched on November 26, 2011. The life cycle cost for the project has increased $881 million since its original baseline in 2008, which includes an 84 percent increase in development costs. The project launched with a risk that the rover's sample analysis drill will short circuit that could cause interference with the avionics systems and limit drill operations. In addition, the project did not complete all of the software for entry, descent, landing, and surface activities, which the project plans to complete during the spacecraft's cruise to Mars. Similarly, the project also plans to close out a number of Problem/Failure Reports identified during development and testing, during the cruise phase.

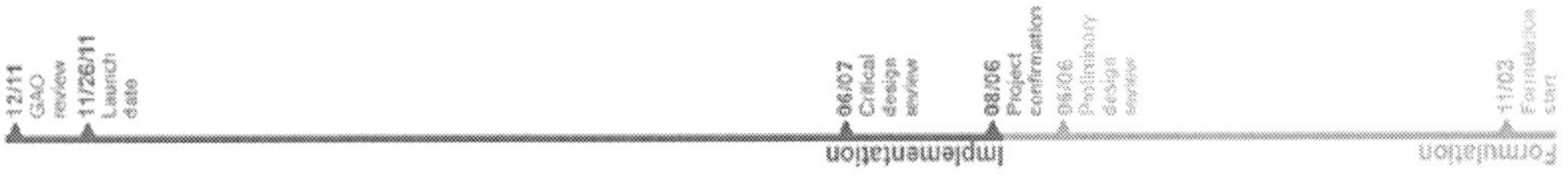

Project Update

MSL successfully launched on November 26, 2011, and is currently en route to Mars, where it is scheduled to land in August 2012. Since the original project baseline in 2008, the life-cycle cost for the project has increased by over $881 million—including an 84 percent increase in development costs—and the launch was delayed from September 2009 until November 2011 since launch windows for Mars missions are optimally aligned every 26 months. These cost and schedule overruns were driven by design and technical problems.

Design Issues

Design issues leading up to the project's launch caused delays in testing and continues to threaten the project's operations. For example, problems with the sample analysis drill, which collects samples from the Martian rocks, were discovered late in the testing and verification of the rover and subsequently caused a rework of internal rover components, delaying the rover's test schedule by 5 weeks. To mitigate this and other problems, the project increased its workforce, which exhausted the project-held cost reserves. While these issues with the sample analysis drill were addressed in the redesign, the project identified additional issues with the drill after the rover had been delivered to Kennedy Space Center for launch processing. The project identified a risk that the sample analysis drill will short circuit to the drill framework—or chassis—when in percussion mode, causing interference from drilling operations that can affect the avionics systems. The project made modifications to the rover that will allow it to detect the short before the avionics hardware is damaged. The project is also able to use the drill in a rotary only mode. This issue could limit the operations of the drill and the project may have to process only softer rock samples, which may significantly increase the risk of not accomplishing mission science objectives.

The project has additional concerns regarding the spacecraft's software that enable its functionality once it arrives at the landing site. Project officials stated that the basic software for landing and traversing exists, but it needs to be upgraded in order to achieve full capability. The project plans to release updates and test its flight software for entry, descent, and landing (EDL) and software for surface operations during the spacecraft's 9-month cruise phase to Mars. The project reported that although this work has slipped into the operations phase after launch, the project believes it has adequate personnel and financial resources to successfully complete EDL, enter surface operations, and complete the mission. NASA headquarters officials, however, stressed the significant amount of work remaining in this area, disruption of which, according to these officials, could impact the project's ability to land safely on Mars. In January 2012, NASA reported that operations costs will increase by almost $68 million, including $8.7 million carried forward from the development phase and an additional $59 million to ensure achievement of mission success criteria and accommodate development of surface mobility flight software.

The project experienced a large number of Problem/Failure Reports (PFR) identified during the development and testing of the rover. For example, the project had over 1,200 PFRs open in early 2011. In June 2011, the NASA Inspector General reported that it found that project managers did not consistently identify and assess the risks associated with the PFRs and, as a result, closing the large number of open PFRs became a point of emphasis for the project during 2011. Project officials said they made it a priority to close those PFRs related to flight and launch operations and then concentrate on those related to software. The project plans to close multiple unresolved software PFRs during the spacecraft's cruise phase. As of October 2011, the project had 392 PFRs still open, although many of those were in review or awaiting signature.

Project Office Comments

The MSL project provided technical comments to a draft of this assessment, which were incorporated as appropriate. Project officials also commented that to date all has gone well with the cruise spacecraft and rover checkout, and that Curiosity is on track for entry, descent, and landing on the Martian surface in 2012.

NPOESS Preparatory Project

Recent/Continuing Project Challenges

- Development Partner Issues

Previously Reported Challenges

- Technology Maturity
- Complexity of Heritage Technology
- Design Stability
- Launch Issues

Project Essentials

NASA Center Lead: **Goddard Space Flight Center**
Partner: **National Oceanic and Atmospheric Administration and U.S. Air Force**
Launch Date: **October 28, 2011**
Launch Location: **Vandenberg AFB, CA**
Launch Vehicle: **Delta II**
Mission Duration: **5 years**
Requirement derived from: **Continuation of Earth Observing System Missions**

Contract Information

Major Contractor: **Northrop Grumman Electronic Systems**
Type of Contract: **Cost Plus Award Fee**
Date of Award: **December 2000**
Initial Value of Contract: **$241.1 million**
Current Value: **$196.9 million**

The National Polar-orbiting Operational Environmental Satellite System (NPOESS) Preparatory Project (NPP) is a joint mission with the National Oceanic and Atmospheric Administration (NOAA) and the United States Air Force. The satellite has five instruments and will measure ozone, atmospheric and sea surface temperatures, land and ocean biological productivity, Earth radiation, and cloud and aerosol properties.

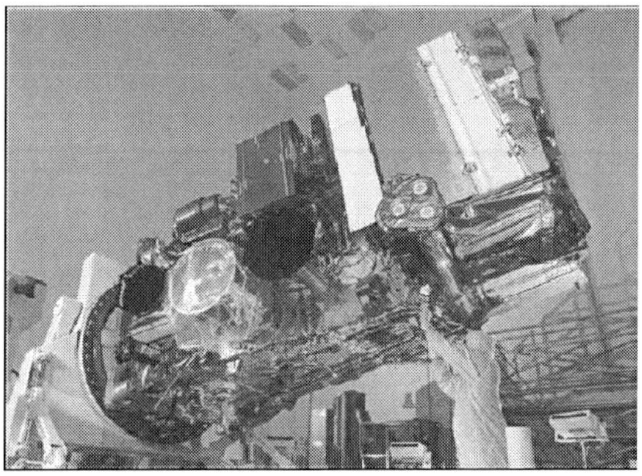

Source: Ball Aerospace.

Project Performance

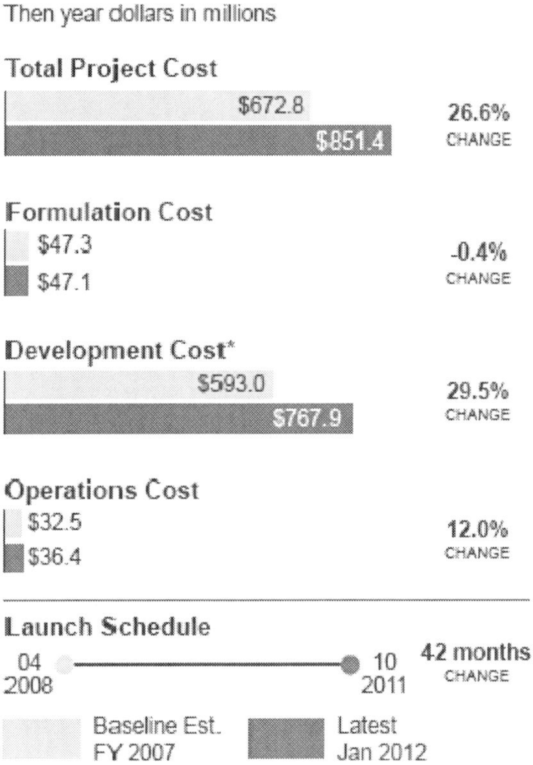

* Represents a 50% growth in development costs since the original baseline of $513 million established in fiscal year 2004.

Project Summary

NPP launched in October 2011, and will provide data for climate research and weather prediction models until future NOAA and DOD satellites are launched and operational. Since 2007, the project experienced almost $175 million in development cost growth and a 42-month launch readiness delay. NPP launched with residual risks on instruments, and work remaining to complete its ground system. Project officials stated that they will continue to perform risk reduction on the ground system and gradually transition to operations in late 2012.

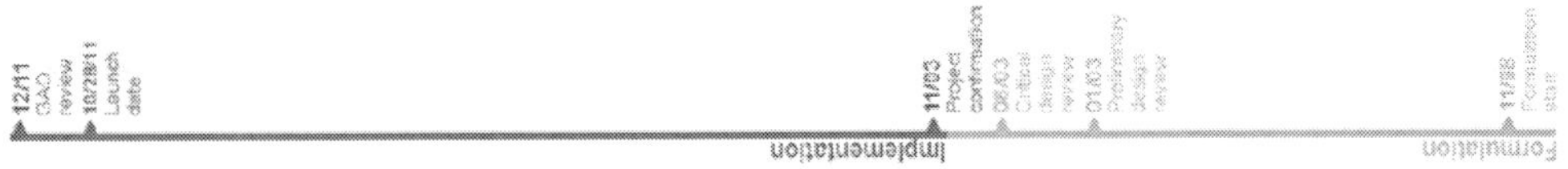

Project Update

INPP successfully launched on October 28, 2011. Originally conceived as a risk reduction mission for the NPOESS mission, NPP will now function as an operational satellite and provide data to be used by the scientific community for studying long-term climate patterns and improving short-term weather forecasts until the launch and operation of future NOAA satellites. As launched, the project was delayed by 42 months and its development costs were almost 37 percent over the project's 2007 baseline; this represents a 58 percent growth in development costs from the project's original 2004 baseline. We have previously reported that NPP project officials attributed cost and schedule overruns to development partner challenges and a lack of central authority between the three NPOESS agencies. In February 2010, the White House announced plans to restructure the NPOESS program. Since that time two new programs were initiated: the NASA-NOAA Joint Polar Satellite System (JPSS) and the DOD Defense Weather Satellite System (DWSS). However, DWSS was subsequently terminated in fiscal year 2012. As part of the NPOESS restructuring, management of the ground system contracts was transferred to the NASA-NOAA JPSS project office.

Development Partner Issues

Project officials stated that one of the most significant obstacles they had to overcome for an October 2011 launch was the development of the ground system network previously overseen and under resourced by the Integrated Program Office (IPO), a joint U.S. Air Force and NOAA program office that managed the NPOESS program. Project officials stated that the ground system network had a lack of resources provided by the IPO when its instruments were behind schedule. Significant work was performed to get the ground system into a launch ready configuration including transferring management of the contracts from the Air Force to the JPSS project office and project officials said they performed major software rebuilds. Work on the ground system to support NPP is not complete. In order to be ready for launch, project officials said they focused on parts of ground system development that were absolute priorities for a launch ready configuration. Project officials stated that after launch, the project will continue to work on the ground system to perform risk reduction and will gradually transition to operations beginning in late 2012. The ground systems team will also continue to evolve the grounds systems to eventually support the JPSS satellites. JPSS is currently planning to launch two satellites, one in 2016 and the other in 2021.

Other Issues to be Monitored

The project launched with residual risk on three partner-provided instruments. NPP project officials stated they lack confidence in the processes used by the IPO for the Visible Infrared Imaging Radiometer Suite, the Cross-track Infrared Sounder, and the Ozone Mapper Profiler Suite instruments. Project officials stated that it is possible that the instruments will not last for the full 5 years of mission duration. We testified in September 2011 that early expiration of the partner-provided instruments would result in a data gap in data coverage, because the first JPSS satellite is not scheduled to launch until 2016.

Project Office Comments

The NPP project provided technical comments to a draft of this assessment, which were incorporated as appropriate. NASA officials added that NPP and the NPP ground system have captured, processed, and distributed over 1 million products since launch and will continue efforts to activate, commission, and calibrate the remaining instruments.

Orbiting Carbon Observatory 2

Recent/Continuing Project Challenges

- Launch Issues
- Funding Issues
- Parts Issues
- Design Issues

Project Essentials

NASA Center Lead: **Jet Propulsion Laboratory**
International Partner: **None**
Projected Launch Date: **February 2013**
Launch Location: **Vandenberg AFB, CA**
Launch Vehicle: **To be determined**
Mission Duration: **2 years**
Requirement derived from: **Earth System Science Pathfinder Announcement of Opportunity 3**

Contract Information

Major Contractor: **Orbital Science Corporation**
Type of Contract: **Cost Plus Fixed Fee/Incentive Fee**
Date of Award: **May 2010**
Initial Value of Contract: **$48 Million**
Current Value: **$48 Million**

NASA's Orbiting Carbon Observatory 2 (OCO-2) is designed to enable more reliable predictions of climate change and is based on the original OCO mission that failed to reach orbit in 2009. It will make precise, time-dependent global measurements of atmospheric carbon dioxide. These measurements will be combined with data from a ground-based network to provide scientists with the information needed to better understand the processes that regulate atmospheric carbon dioxide and its role in the carbon cycle.

Source: Jet Propulsion Laboratory (artist depiction).

NASA expects enhanced understanding of the carbon cycle will improve predictions of future atmospheric carbon dioxide increases and the potential impact on the climate.

Project Performance

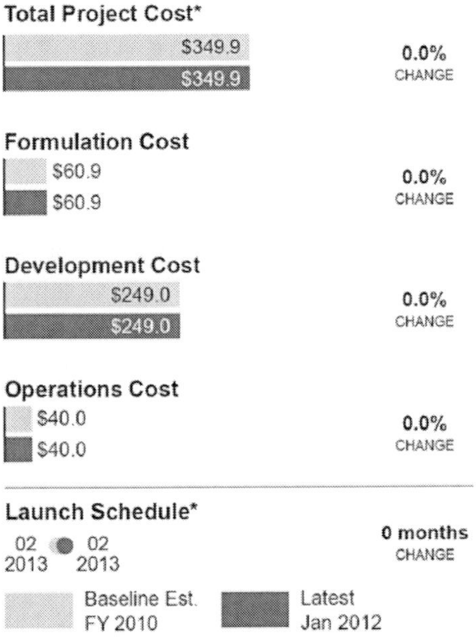

* The project's cost and schedule are currently under review pending resolution of launch vehicle issues.

Project Summary

NASA had selected the Taurus XL launch vehicle for OCO-2, the same vehicle used for the OCO and Glory missions. After the failure of the Taurus XL to place either spacecraft into orbit, NASA stopped payment for the Taurus XL and is reevaluating launch vehicle options for OCO-2. Prior to the Glory failure, the project appeared to be on schedule, but launch vehicle uncertainties will delay the OCO-2 launch date. Despite being a rebuild of the original OCO mission, costs have increased due to parts obsolescence, purchase of spares, launch vehicle cost increase, and inflation. Similar to the OCO mission, the project is experiencing an issue with spectrometer motion.

Project Update

Launch Vehicle Issues

OCO-2 has been designed to launch on a Taurus XL, the same launch vehicle as was used for OCO; however, Taurus XL has failed to place both the OCO and Glory missions into orbit. As a result, the current launch vehicle for OCO-2 is "to be determined" and work under the Taurus XL task order has been suspended. The project reports that the schedule for OCO-2 has already been adversely affected as there are no viable launch vehicle alternatives that preserve the February 2013 launch date. The project expects a major replanning effort will be necessary and the technical and cost implications are unknown. Potential alternative launch vehicles currently available under the NASA Launch Services-II contract for OCO-2 include the Falcon 9 and the Delta II. According to NASA launch services officials, it is likely the earliest OCO-2 could be launched would be at least 2 years from the selection of an alternate launch vehicle. In January 2012, NASA reported that the project's life cycle cost estimate and schedule were under review pending resolution of these launch vehicle issues.

Funding Issues

In December 2008, OCO's life-cycle cost estimate was $273.1 million, compared to OCO-2's 2010 baseline estimate of $349.9 million. Project officials attributed the higher life-cycle cost estimate for OCO-2 due to an increase in reserves—or unallocated future expenses—held by the Science Mission Directorate, an increase of over $20 million for the Taurus XL launch vehicle, an increase of $25 million due to inflation, and the need for $10 million to procure a new cryocooler. The life-cycle cost for OCO-2 may increase once a launch vehicle decision is made since some of the alternatives are more expensive than that of the Taurus XL. In addition, a launch delay may necessitate retention of personnel for the extended period, which could be costly.

Parts Issues

The project is making every effort to duplicate the original OCO design using identical hardware, drawings, documents, procedures, and software wherever possible and practical in order to produce OCO-2 with minimum cost, schedule, and technical risk. However, project officials stated that there were no engineering models for many of the OCO components and the original components were lost when OCO failed to reach orbit. The OCO-2 project will procure a full set of spares to help avoid problems with further parts obsolescence during the development and testing of flight hardware. The project has encountered difficulties with particular components. For example, OCO-2 had to redesign the flight computer to accept a new memory chip in order to avoid converting the entire spacecraft for a new flight computer. The suppliers of OCO's X-band and S-band receivers went out of business and ceased production, respectively, and these will have to be replaced or redesigned. The project has also had to use 2 months of schedule reserve to remove and reapply metallic plating that was flaking off after improper application by a subcontractor.

Project Office Comments

The OCO-2 project provided technical comments to a draft of this assessment, which were incorporated as appropriate. Project officials also commented that the project's philosophy to take advantage of existing OCO designs, plans, and procedures has served to reduce risk for the most part. They added that there have been instances where contracting with the same vendor and applying the same processes followed on OCO prevented planned results from being achieved, and that a "build-to-print" approach is no guarantee of initial quality and may require extensive rework.

Design Issues

OCO-2 includes a single instrument, the three-channel grating spectrometer, based on heritage technology from OCO. The instrument on OCO-2 had a spectrometer stability issue similar to that experienced on OCO, which may reduce the amount of useable data that the instrument is able to collect. The project thought that it had fixed the problem on OCO after a successful test, but it has re-emerged on OCO-2. Project officials report that a stabilizer was added to each of the three spectrometer channels to mitigate further motion. They added that test results are promising; however, final verification is pending. Due to its heritage design, the project had released 95 percent of its engineering drawings at the critical design review.

Orion Multi-Purpose Crew Vehicle

Project Challenges

- Funding Issues
- Design Issues

Project Essentials

NASA Center Lead: **Johnson Space Center**
Partner: **None**
Projected First Non-Crew Launch Date: **2017**
Projected First Crew Launch Date: **2021**
Launch Location: **Kennedy Space Center, FL**
Launch Vehicle: **Space Launch System**
Mission Duration: **Varied based on destination**
Requirement derived from: **NASA Authorization Act of 2010**

Contract Information

Major Contractor: **Lockheed Martin**
Type of Contract: **N/A**
Date of Award: **N/A**
Initial Value of Contract: **N/A**
Current Value: **N/A**

Project Performance

Then year dollars in millions

Latest: Jan 2012 **Not available**

* The project has not yet reached the point in the acquisition life cycle where a preliminary life-cycle cost estimate would normally be developed.

Preliminary estimate of Project Life Cycle Cost*

Launch Schedule	**2017 First non-crew**
	2021 First crew

NASA's Orion Multi-Purpose Crew Vehicle (MPCV) is a crew vehicle being developed to conduct in-space operations beyond low Earth orbit. The current design for MPCV consists of a crew module, service module, and launch abort system. The MPCV spacecraft will provide for safe crew transportation and habitability during the ascent, in-space operations and entry, descent, and landing mission phases. MPCV is planned to eventually launch atop NASA's Space Launch System to conduct exploration missions beyond low earth orbit and can be utilized to service the International Space Station (ISS) if necessary.

Source: Lockheed Martin Space Systems Company.

Project Summary

The MPCV program will continue to develop systems of the former Orion project. NASA has returned MPCV to Phase A, even though the program continues to conduct activities normally performed later in formulation. Program officials stated that, given budget constraints, the first MPCV non-crewed flight is scheduled for 2017 and the first crewed flight for 2021. An independent assessment of initial budget estimates for MPCV found the estimates reasonable for near-term planning, but inadequate for the development of a program baseline and that program reserves are insufficient. Currently, 15 missions are being considered for MPCV, and as such the program currently does not have a specific mission to guide its design. The program office reports that each of the 15 missions could be accomplished using the core MPCV capabilities.

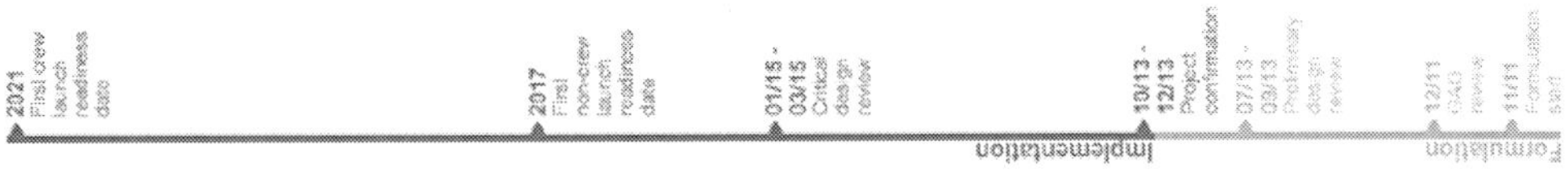

Project Update

Thc NASA Authorization Act of 2010 directed NASA to develop an MPCV that will continue to advance the development of the human safety features, designs, and systems of the former Orion project. To transition from Orion to MPCV, NASA has traced the current requirements for MPCV to the former Orion plan and contracts to ensure that the program is meeting requirements of the Authorization Act and utilizing previous contracts to the extent practicable. Although NASA reported spending $4.9 billion through November 2010 on Orion, which was in Phase B, NASA has placed MPCV in the concept and technology development phase of formulation—Phase A. According to NASA officials, this placement was necessary due to continued work on refining budget estimations and aligning MPCV requirements with the other associated and newly created programs—the Space Launch System (SLS) and the 21st Century Ground System. MPCV, however, is continuing development work and testing to demonstrate key technologies started during the former Orion project that, according to program officials, is normally conducted in later phases of development. For example, program officials stated that water landing tests were conducted on the crew module in July 2011, and acoustic and vibration tests of the capsule were conducted in September 2011.

Funding Issues

NASA officials stated that the budget for the program is more constrained than was planned for Orion; therefore, the first non-crewed flight is not scheduled until 2017 and the first crewed flight in 2021, as compared to initial operational capability planned for 2015 under Orion. An independent assessment of initial budget estimates for the MPCV found the estimates reasonable for short-term planning, but inadequate for the development of a program baseline. Further, given that NASA had not conducted a risk analysis, the report stated that program reserves are insufficient. According to NASA officials, the agency is conducting a comprehensive review in preparation for the President's fiscal year 2013 budget proposal in February 2012, but will not be able to provide a baseline cost estimate for MPCV until February 2013 when NASA expects to have greater clarity of its alignment to the associated programs.

Program officials said NASA modified the former Orion prime contract with Lockheed Martin to ensure that for MPCV, spacecraft requirements were phased to match the anticipated budget allocation. The program intends to include an early test flight—designated Exploration Flight Test 1—scheduled for early 2014. Program officials reported this test flight should allow them to address 10 out of 13 top risks related to loss of crew, such as the capability of the heat shield. To conduct this test, Lockheed Martin will need to acquire a launch vehicle since SLS will not be available. NASA is negotiating with Lockheed Martin to determine how the additional test-unique costs will be accommodated in the existing contract.

Design Issues

The program does not have a specific mission to guide its design. NASA is currently considering 15 potential missions, ranging from non-International Space Station (ISS) low earth orbit utilizations, to lunar surface operations, to landing on Mars. NASA officials did not know when the missions MPCV will conduct will be determined, as they stated these decisions are typically made by the Presidential Administration and Office of Management and Budget. As MPCV is intended to be a multipurpose vehicle, the program office reports that each of the 15 missions could be accomplished using the core MPCV capabilities. Program officials said one area of lower priority work being deferred is the acquisition of components for MPCV needed to enable it to support ISS. They said, however, that supporting ISS is being taken into consideration during design of MPCV and integrating these capabilities to the spacecraft would take approximately 2 years, should the need arise.

PROJECT OFFICE COMMENTS

The Orion MPCV program office provided technical comments to a draft of this assessment, which were incorporated as appropriate. Program officials also commented that the program is following the 2010 NASA Authorization Act and agency authorization to develop the spacecraft using existing contracts as applicable.

Radiation Belt Storm Probes

Recent/Continuing Project Challenges

- Design Issues
- Parts Issues
- Contractor Issues

Project Essentials

NASA Center Lead: **Goddard Space Flight Center**
Partner: **National Reconnaissance Office**
Projected Launch Date: **September 2012**
Launch Location: **Cape Canaveral AFS, FL**
Launch Vehicle: **Atlas V**
Mission Duration: **2 years**
Requirement derived from: **Solar and Space Physics Decadal Survey, 2003**

Contract Information

Major Contractor: **Johns Hopkins University/Applied Physics Laboratory**
Type of Contract: **Cost Plus Fixed Fee**
Date of Award: **2006**
Initial Value of Contract: **$435.5 million**
Current Value: **$502.4 million**

Project Performance

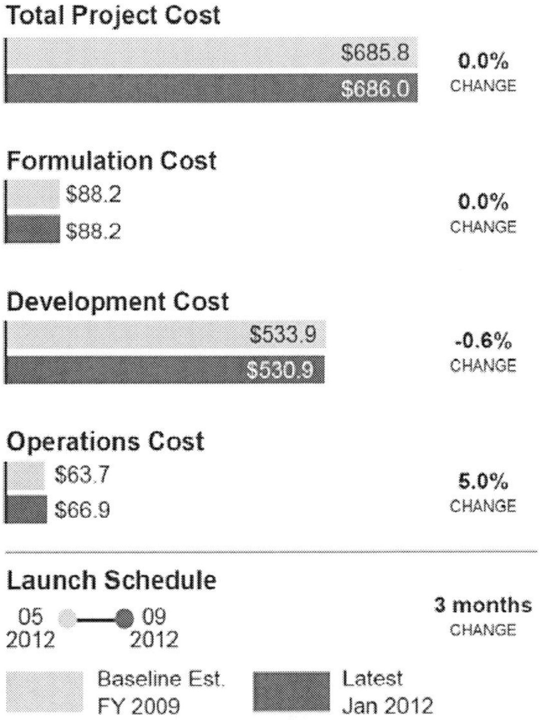

Source: © 2010 The Johns Hopkins University/Applied Physics Laboratory (artist depiction). All Rights Reserved.

The Radiation Belt Storm Probes (RBSP) mission will explore the Sun's influence on the Earth and near-Earth space by studying the planet's radiation belts at various scales of space and time. This insight into the physical dynamics of the Earth's radiation belts will provide scientists with data to make predictions of changes in this little understood region of space. Understanding the radiation belt environment has practical applications in the areas of spacecraft system design, mission planning, spacecraft operations, and astronaut safety. The two spacecraft will measure the particles, magnetic and electric fields, and waves that fill geospace and provide new knowledge on the dynamics and extremes of the radiation belts.

Project Summary

The RBSP project reported in July 2011 that it was approved for a replan that delays the launch readiness date 3 months to September 2012 because of conflicting launch schedules of other missions. However, the project had low schedule and cost reserves prior to the replan due to late delivery of instruments. In particular, manufacturer revisions to the design of a component for the Helium-Oxygen Proton-Electron (HOPE) instrument caused a slip in schedule because the component was not ready to be incorporated into the instrument. In addition, a chiller unit failure during winter storms and facility shutdowns during recent forest fires at Los Alamos National Laboratory caused additional delays for the HOPE instrument.

Project Update

In July 2011, the RBSP project reported it was approved for a replan that will delay the launch readiness date from May 2012 to September 2012. According to project officials, the project will not exceed its current life-cycle cost estimate due to this delay. Project officials said that the replan was determined necessary by NASA because of changes to the launch manifest caused by other missions, including some from the Department of Defense. The project manager stated that there will be some cost impact to keep personnel on the project longer, but that NASA will likely release headquarters' reserves because such a delay was beyond the control of the project. The project reported that in April 2011, the schedule margin had burned down to 22 days prior to the announcement of the replan, even though the project had planned for 42 days of schedule reserve at that point. The program manager at the Applied Physics Laboratory also stated that there were actions, such as extended shifts, that could have been taken to maintain the May 2012 launch date within existing project reserves if the replan had not been implemented.

Design/Parts Issues

RBSP project officials told us that they had to make adjustments to the integration and test schedule, such as adding more weekend shifts, to accommodate late deliveries of several instruments because of design and parts related issues. They added that having two spacecrafts on the mission allows for flexibility in the schedule to mitigate the effects of late instrument deliveries. In the spring of 2011, the project had been experiencing low cost and schedule reserves mostly due to instrument delays, including one involving the high voltage optocoupler on the Helium-Oxygen-Proton-Electron (HOPE) instrument. The manufacturer made revisions to the optocoupler design and NASA reported that other projects experienced issues with similar parts. The project reported that the redesign will require new high-voltage power supply boards for the new optocouplers, and those boards are estimated to cost approximately $900,000—an amount covered by the project's reserves, according to project officials. The project proceeded with the delivery of the HOPE instrument while the new boards were being built with the new optocouplers. The project is awaiting the delivery of the new parts to integrate into the instrument, a process officials except to happen in March 2012. During part-level testing the project discovered a failure in the qualification unit of the

Magnetic Electron Ion Spectrometer (MagEIS) instrument. The project reported that, as of August 2011, the failure analysis to date has been inconclusive, and the root cause is indeterminate. Project officials, however, believe the problem was caused by electro-static discharge during handling. As of August 2011, five of the eight MagEIS instruments were yet to be delivered. In addition, the project reported a delay in the delivery date of the transceiver subsystem to integration and test after the vendor reported problems with its filters during electrical testing. The project also experienced instrument delivery delays with the Electric Fields and Waves instrument after anomalies were detected during thermal vacuum testing.

Contractor Issues

In addition to technical issues with the HOPE instrument, some of the delivery delays for the HOPE instrument were due to events outside the control of the project. Two separate events at the Los Alamos National Laboratory—a chiller unit failure during storms last winter and a facility closure during forest fires in 2011—caused about a 2-month delay for that instrument. During a design review for a Power System Electronics (PSE) box, the project also discovered that a vendor had made changes to how its regulators are used. Project officials reported that they had found out about the changes through an independent reviewer, not the vendor. As a result of the changes, the RBSP project reviewed all applications of the vendor's regulators used for the RBSP project and determined it needed to make modifications to three flight components, including the PSE box.

Project Office Comments

The RBSP project office provided technical comments to a draft of this assessment, which were incorporated as appropriate.

Soil Moisture Active and Passive

Recent/Continuing Project Challenges

- Launch Issues
- Funding Issues
- Technology Issues

Project Essentials

NASA Center Lead: **Jet Propulsion Lab**
Partner: **None**
Projected Launch Date: **October 2014 – January 2015**
Launch Location: **TBD**
Launch Vehicle: **TBD**
Mission Duration: **3 years**
Requirement derived from: **2007 Earth Science Decadal Survey**

Contract Information

Major Contractor: **Northrop Grumman Aerospace Systems**

Type of Contract: **Cost Plus Fixed Fee**

Date of Award: **June 2009**

Initial Value of Contract: **$18.56 Million**

Current Value: **$27.50 Million**

Project Performance

Then year dollars in millions

* This estimate is preliminary, as the project is in formulation and there is uncertainty regarding the costs associated with the design options being explored. NASA uses these estimates for planning purposes.

Preliminary estimate of Project Life Cycle Cost*

Launch Schedule 10/2014 – 01/2015

Source: Jet Propulsion Laboratory (artist depiction).

NASA's Soil Moisture Active and Passive (SMAP) is one of four first-tier missions recommended by the National Research Council's 2007 Earth Science Decadal Survey. SMAP leverages previous Earth Science missions and is based on the soil moisture and freeze/thaw mission concept developed by an earlier mission known as Hydros. The SMAP mission will provide new information on global soil moisture and its freeze/thaw state enabling new advances in hydrospheric science and applications. The measures will improve understanding of regional and global water cycles, improve weather forecasts, flood and drought forecasts, and climate changes.

Project Summary

SMAP held its preliminary design review 7 months late due to a delay in launch vehicle selection. The project is proceeding with the assumption it will use a Minotaur IV-plus launch vehicle; however, final launch vehicle selection is still pending. While SMAP is utilizing heritage technologies to reduce development risks, none of the technologies were mature or were planned to be matured at the preliminary design review. The project is working with the federal agencies regarding the potential for SMAP's radar to interfere with air surveillance and Global Positioning System (GPS) frequencies. The project has submitted a mitigation plan to the Federal Aviation Administration (FAA) and initial tests indicate that the project does not cause interference.

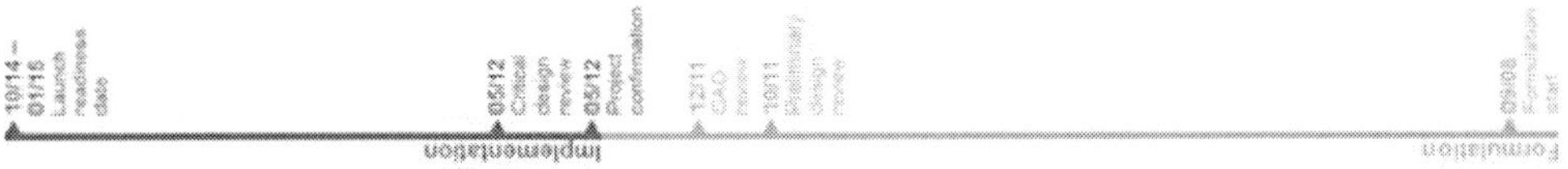

Project Update

Launch Issues/Funding Issues

The launch vehicle selection for the SMAP project did not occur at the project's preliminary design review, which had already been delayed by 7 months as a result of the launch vehicle issue. The project is proceeding with the assumption that it will launch on a Minotaur IV-plus and is currently working with the Air Force and Orbital Sciences Corporation, the contractor for Minotaur, on a coupled loads analysis for the that vehicle to understand the environment for the spacecraft during launch.

The final decision on selection of a launch vehicle for SMAP is pending the results of the Request for Launch Services Proposal that was released for industry comment in early 2012. If a different launch vehicle is selected, the project will likely experience cost increases since the vehicles offered under the current launch services contract would be significantly more expensive for the project than Minotaur. While continuing to design to accommodate multiple launch vehicles is possible, the project manager said that it limits design capabilities and can raise costs to the project.

Issues surrounding the launch vehicle selection have resulted in NASA delaying the project's confirmation review by six months until May 2012 and increasing the lower limit of the project's life-cycle cost range estimate by over $90 million.

Technology Issues

The project held its mission preliminary design review in October 2011 without maturing any of its critical or heritage technologies. The project has three heritage technologies—the radar, radiometer, and the reflector boom assembly—all of which have been used on other missions and will need to be adapted for use on the SMAP project. The project is tracking the radiometer as a risk since it requires additional filtering to lessen radio frequency interference. The project has identified the spectral filtering as a critical technology, but this technology was not mature at the mission preliminary design review. The project expects to qualify all of the technologies for use prior to the mission critical design review in May 2012.

Other Issues to be Monitored

The project is working with the FAA and the National Telecommunications and Information Administration (NTIA) to acquire spectrum approval because there is the potential for SMAP's radar to interfere with air surveillance and GPS frequencies. If the NTIA does not approve of the project's mitigation plan, the result could be a loss of science measurements. The project mitigation plan is to use frequency hopping—or changing the frequency every 12 seconds—to prevent interference. According to NASA, initial tests indicate that this mitigation is effective to reduce potential interference to levels that have traditionally been acceptable to the FAA. The project has also had delays with flight software development due to flight software resources being constrained by other projects at the Jet Propulsion Laboratory. The project is currently developing a recovery plan to mitigate the impact of the delay.

The project is working with the European Space Agency's Soil Moisture and Ocean Salinity project as well as the Aquarius project to help develop a radio frequency interference map for SMAP. As the spacecraft orbits the earth, it will encounter radio frequency interference that varies by location and new sources of radio frequencies as the environment is not static. The radio frequency interference map will allow the project to select radio frequencies with less interference for gathering data depending on where the spacecraft is and what interference is expected in that area.

Project Office Comments

The SMAP project provided technical comments to a draft of this assessment, which were incorporated as appropriate. Project officials also commented that SMAP has no technology issues since the project uses a test approach where formal qualification is performed on flight hardware. They added that engineering models are being developed for adapted designs.

Solar Probe Plus

Recent/Continuing Project Challenges

- Launch Issues

Project Essentials

NASA Center Lead: **Goddard Space Flight Center**
International Partner: **None**
Projected Launch Date: **July 2018**
Launch Location: **Eastern Range**
Launch Vehicle: **TBD**
Mission Duration: **7 years**
Requirement derived from: **Solar and Space Physics Decadal Survey, 2003**

Contract Information

Major Contractor: **Aerospace Research Development & Engineering Support**
Type of Contract: **Cost Plus Award Fee**
Date of Award: **May 2010**
Initial Value of Contract: **$38.45 Million**
Current Value: **$52.58 Million**

Project Performance

Then year dollars in millions

* This estimate is preliminary, as the project is in formulation and there is uncertainty regarding the costs associated with the design options being explored. NASA uses these estimates for planning purposes.

Preliminary estimate of Project Life Cycle Cost*

Solar Probe Plus (SPP) will explore the Sun's outer atmosphere, or corona, as it extends into space. The spacecraft will orbit the Sun 24 times and its instruments will observe the generation and flow of solar wind from very close range. By observing the corona, where solar energetic particles are energized, there is potential to further science in terms of shedding light on two central issues of heliophysics: the origin and evolution of solar wind and why the sun's outer atmosphere is so much hotter than the visible surface. In order to achieve its mission, parts of the spacecraft must be able to withstand temperatures exceeding 2,500 degrees Fahrenheit as well as endure blasts of extreme radiation.

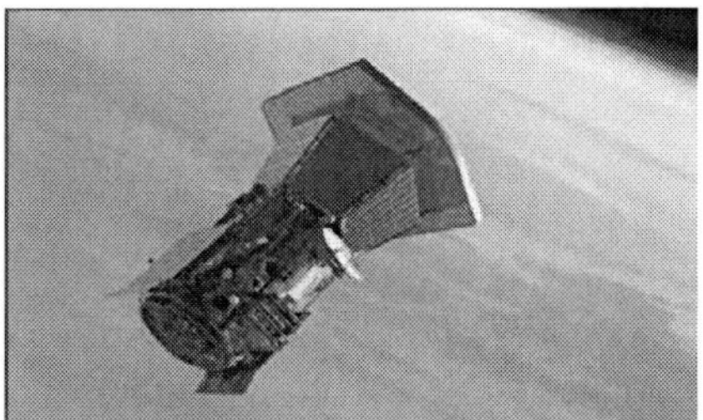

Source: © 2010 Johns Hopkins University/Applied Physics Laboratory (artist depiction).

Project Summary

The Solar Probe Plus entered the preliminary design and technology completion phase of formulation in January 2012 with a preliminary life cycle cost range estimate of $1.233 billion to $1.439 billion. The project rescheduled key project reviews from May 2011 to November 2011 to allow it to further define the science requirements and perform technical trades to choose the final spacecraft configuration. Chief risks to the project in terms of cost and schedule include developing a sunshield capable of protecting the instruments from the harsh near-Sun environment, developing a cooling system for the retractable solar array panels, and achieving the total launch energy necessary to launch the spacecraft toward the Sun.

Project Update

The SPP project mission definition review (MDR) and system requirements review (SRR) were scheduled for May 2011, but were rescheduled for November 2011 to allow the project to further define the science requirements and perform technical trades to choose the final spacecraft configuration and submit preliminary cost and schedule data to NASA headquarters. Delay of these reviews has led to a eight-month slip to its agency-level review to approve the project to move into Phase B—the preliminary design and technology completion phase—until January 2012, when it reported a preliminary life cycle cost range estimate of $1.233 billion to $1.439 billion.

Launch Issues

SPP project officials reported that they are designing the project to launch on the Atlas V, which is currently the only vehicle offered under the NASA launch services contract (NLS II) that can support the project's launch energy. The other launch vehicles NASA has studied are the Delta IV Heavy and the Falcon Heavy. Although these vehicles are not currently offered under the NLS II contract, they are being considered because there is a potential that they could be offered under the contract in time for use by SPP. According to project officials, regardless of the vehicle selected, an additional upper stage rocket will be required because one of the mission's key challenges is achieving the total launch energy necessary to launch the spacecraft toward its ultimate destination—an orbit around the Sun. Because of the high launch energy required for SPP, the project has to maintain strict mass management to take full advantage of the launch vehicle capability. NASA has determined that the Atlas V can support the SPP project's mass allocation with a new solid rocket motor upper stage that can be modified from other existing upper stage technologies.

Other Issues to be Monitored

A key challenge of the SPP mission will be the development of critical technologies that will allow science instruments to function within the harsh near-Sun environment. Project officials stated the spacecraft will take measurements at about 4 million miles from the surface of the Sun—closer than any previous spacecraft. In particular, project officials reported that they are concerned with their ability to build and test the Thermal Protection System (TPS)—a carbon-foam filled sun shield that would shield the instruments from the direct heat and radiation of the Sun—at full scale. Officials expect to develop and mature a full prototype of the TPS for testing during the project's preliminary design phase.

Project officials are also focused on the development and production of two sets of solar arrays—essentially solar power generators—that will retract and extend as the spacecraft moves toward or away from the Sun. The solar array cooling system is being developed to ensure the solar panels stay at required temperatures. A key to developing the cooling system for the solar arrays, is the requirement for it to dissipate up to 5,000 watts of thermal energy during the spacecraft's closest approach to the Sun. The project has plans to fabricate and test a solar array and a half-scale cooling system as well as a full sized actively cooled secondary section solar array prior to the mission preliminary design review.

Although the key technologies will be tested in environments similar to mission conditions, it will be impossible to replicate the extreme conditions the fully assembled probe will be exposed to during its closest proximity to the Sun. Therefore, functionality of the entire spacecraft in the near-Sun environment cannot be verified fully by testing prior to launch. As a result, project officials said they plan to use simulators for the TPS and Solar Arrays in systems test. Project officials said they plan to test all of the elements of the spacecraft with a heat simulator and then model against the data from these tests.

PROJECT OFFICE COMMENTS

The SPP project provided technical comments to a draft of this assessment, which were incorporated as appropriate. Project officials also commented that they believe the project remains on track for a July 2018 launch readiness date with full schedule reserve.

Space Launch System

Project Challenges

- Funding Issues
- Design Issues

Project Essentials

NASA Center Lead: **Marshall Space Flight Center**
Partner: **None**
Projected First Non-Crew Launch Date: **2017**
Projected First Crew Launch Date: **2021**
Launch Location: **Kennedy Space Center, FL**
Launch Vehicle: **N/A**
Mission Duration: **Varied based on destination**
Requirement derived from: **NASA Authorization Act of 2010**

Contract Information

Major Contractor: **N/A**
Type of Contract: **N/A**
Date of Award: **N/A**
Initial Value of Contract: **N/A**
Current Value: **N/A**

Project Performance

Then year dollars in millions

Latest: Jan 2012 **Not available**

* The project has not yet reached the point in the acquisition life cycle where a preliminary life-cycle cost estimate would normally be developed.

Preliminary estimate of Project Life Cycle Cost*

Launch Schedule **2017 First non-crew**
2021 First crew

The Space Launch System (SLS) is intended to be the Nation's first heavy-lift launch vehicle since the Saturn V was developed for the Apollo program. SLS is planned to eventually launch NASA's Multi-Purpose Crew Vehicle to conduct exploration missions beyond Low Earth Orbit and service the International Space Station if necessary. The vehicle is planned with an initial lift capacity of 70 metric tons to low-Earth orbit and evolvable to 130 metric tons. The initial 70 metric ton capability will include a core stage and two five-segment boosters. The 130 metric ton capability will include a core stage, an upper stage powered by a J-2X engine, and advanced boosters.

Source: SLS Project Office (artist depiction).

Project Summary

The SLS program formally entered into formulation in November 2011. Program officials stated that, given budget constraints, the first SLS non-crewed flight could not be scheduled before 2017 and the first crewed flight for 2021. An independent assessment of initial budget estimates for the SLS program found them to be inadequate for the development of program baselines and stated that reserves are insufficient. In addition, 15 potential missions are being considered for SLS, and as such the program does not have a specific mission to guide its design. Program officials report that these 15 missions are currently being evaluated for SLS to establish the overall technical performance requirements for the vehicle.

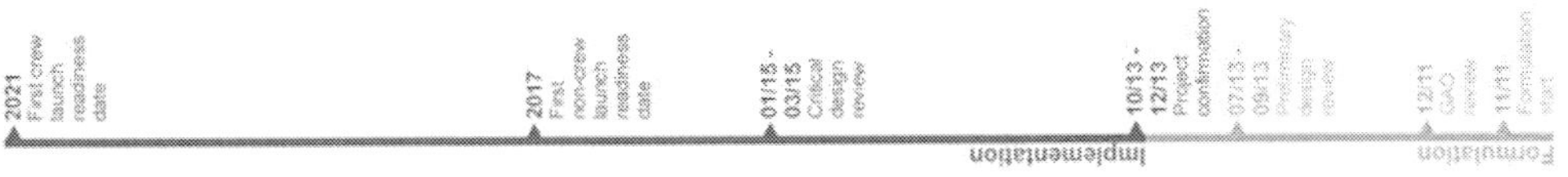

Project Update

The NASA Authorization Act of 2010 directed NASA to develop SLS to leverage investments made in the Ares project and Space Shuttle program and will provide a launch system to Low Earth Orbit and beyond by 2016. Although the Authorization Act required NASA to report to Congress within 90 days of enactment of the Act on the designs, among other things, for the SLS, the agency delayed for several months in order to perform cost estimates and study alternative architectures before issuing its report in September 2011. The program entered formulation in November 2011. To ensure compliance with the Act, NASA is continuing to review investments and workforce from the Shuttle program and the Ares project that can be leveraged for SLS. Program officials reported that the work conducted in fiscal year 2011, originally planned for Ares, remained within the scope of the Ares prime contracts. This allowed work to continue as NASA refined the program's acquisition strategy and made decisions on whether to continue, terminate, or modify the Ares contracts. For example, the program has continued development and testing of the five-segment rocket booster and J-2X engine that will be used for SLS. Although most work is currently being performed through existing contracts, some components, such as the Advanced Boosters will be open to competition in the future. Finally, program officials reported that they are working to identify available vendors for Shuttle-related components since NASA has not recently procured some of these components. For example, 30 major suppliers for external tank components have not been used by NASA since about 2006.

Funding Issues

Although the Authorization Act established a goal to provide SLS capability by 2016, NASA officials stated that the projected funding levels and the time needed to complete all design requirements will not allow the SLS program to achieve full operational capability by 2016. Currently, NASA is targeting 2017 for the first non-crew flight and 2021 for the first crewed flight. Even with the delay to 2017, NASA officials emphasized the need to implement a more cost effective and efficient way of doing business than was in place on the Ares project. These officials specifically cited the importance of having sufficient funding reserves both within the program and at headquarters. An independent assessment of initial budget estimates for the SLS found that although the estimates were reasonable for near-term planning, they are inadequate for the development of a program baseline and stated that program reserves are insufficient. NASA officials stated that the agency is conducting a comprehensive review of SLS cost estimates in preparation for the President's fiscal year 2013 budget request, which will be delivered to Congress in early 2012. The agency, however, will not provide Congress with a baseline life-cycle cost estimate for SLS until February 2013 following the SLS confirmation review, when according to NASA officials, the agency will have greater understanding of SLS costs and their alignment to the associated programs

Design issues

Though two flight tests are planned, the SLS program currently does not have a mission to guide its design. NASA is considering 15 potential missions, ranging from non-International Space Station lower earth orbit utilizations, to lunar surface operations, to landing on Mars. Program officials report that these missions are being evaluated for SLS to establish the overall technical performance requirements for the vehicle. NASA officials did not know when SLS missions will be determined, stating that these decisions are typically made by the Presidential Administration and Office of Management and Budget. SLS program officials stated that the lack of a defined mission is a challenge when trying to design and build a vehicle, because the program will have to build flexibilities into the design to accommodate mission specific requirements.

PROJECT OFFICE COMMENTS

The SLS program office provided technical comments to a draft of this assessment, which were incorporated as appropriate. Program officials also commented that NASA's commitment to a first launch in 2017 is driven both by resource availability and the time needed to complete all design requirements.

Stratospheric Observatory for Infrared Astronomy

Recent/Continuing Project Challenges

- Funding Issues
- Design Issues

Previously Reported Challenges

- Technology Issues
- Contractor Issues

Project Essentials

NASA Center Lead: **Dryden Flight Research Center**
International Partner: **German Space Agency**
Projected Full Operational Capability: **Dec. 2014**
Aircraft: **Modified 747SP**
Sortie Location: **Dryden Aircraft Operations Center, Calif.**
Mission Duration: **20 years of science mission flights**
Requirement derived from: **Astronomy and Astrophysics Committee, National Research Council, 1991**

Contract Information

Major Contractor: **Universities Space Research Association**
Type of Contract: **Cost Plus Fixed Fee**

Date of Award: **December 1996**
Initial Value of Contract: **$484 Million**
Current Value: **$573 Million**

Project Performance

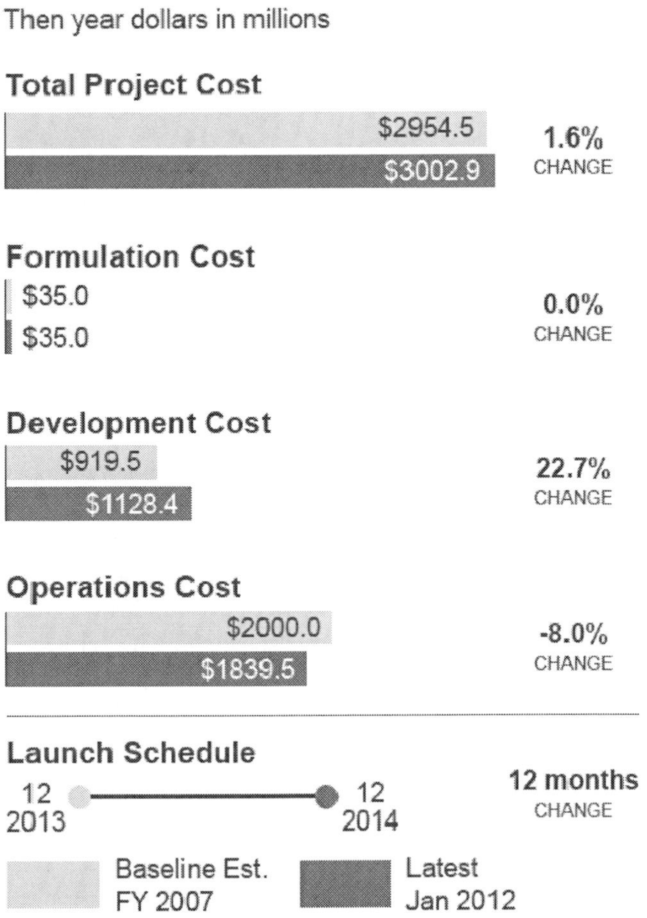

SOFIA is a joint project between NASA and the German Space Agency to install a 2.5 meter telescope in a specially modified Boeing 747SP aircraft. This airborne observatory is designed to provide routine access to the visual, infrared, far-infrared, and sub-millimeter parts of the electromagnetic spectrum.

Its mission objectives include studying many different kinds of astronomical objects and phenomena, including star birth and death; the formation of new solar systems; planets, comets, and asteroids in our solar system; and black holes at the center of galaxies. Interchangeable instruments for the observatory are being developed to allow a range of scientific measurement to be taken by SOFIA.

Source: E. Zavala.

Project Summary

SOFIA has completed basic science flights and achieved its initial operating capability. The project, however, has modified its definition for full operating capability based on conflicting baseline information from 2007. The project will now satisfy full operating capability with only four instruments instead of the eight previously reported. In addition, the project is concerned about potential budget shortfalls in 2014 and beyond and has investigated options to accommodate reduced funding. The project also completed development of the cavity door software that previously caused a one year delay to initial science flights. SOFIA's development costs have increased more than 268 percent, over $1.1 billion, since its 1995 estimate.

Project Update

In December 2010, SOFIA achieved its initial operating capability. The project completed early science flights with the Faint Object Infrared Camera for the SOFIA Telescope (FORCAST) and the German Receiver for Astronomy at Terahertz Frequencies (GREAT) instruments. A total of 13 science flights with FORCAST and 15 science flights with GREAT were completed. In doing so, the project met or exceeded all early science requirements, including those for the telescope elevation and image size. To achieve a full operating capability, the project previously reported that it planned to fly eight instruments on SOFIA. The project, however, has modified its definition of full operating capability based on its recent interpretation of conflicting baseline information contained in the project's 2007 Major Program Annual Report. As a result, the project is currently working toward flying with four instruments to satisfy its 2007 agreement of full operational capability, which the project believes can be achieved prior to December 2014. Two instruments have been successfully integrated with the aircraft and two instruments are in the process of being integrated. Three other instruments are still in development and will be integrated with the aircraft after full operational capability is achieved. According to project officials, one instrument—the Caltech Submillimeter Interstellar Medium Investigations Receiver—was descoped when the project realized in the spring of 2010 that its capabilities were already being performed by the Herschel mission.

Funding Issues

The SOFIA project completed its Joint Cost and Schedule Confidence Level (JCL) process in August 2010. Through this process, the project was estimated to cost $27.7 million more than the estimates prior to the JCL. According to project officials, the JCL included a higher cost of developing the project's instruments and infrastructure improvements at the Dryden Aircraft Operations Facility. The project is concerned about available funding past fiscal year 2014 and has investigated areas of potential compromises if insufficient funds are available to maintain the project's original plans. Overall, SOFIA's development costs have increased more than 268 percent, over $1.1 billion, since its 1995 estimate. As we have reported previously, this increase is partly due to challenges with modification of the aircraft to be used for SOFIA, development of the Cavity Door Drive System (CDDS), and increased flight hanger costs.

Design Issues

The project identified damage to the observatory's primary mirror and the issue was investigated. Initial analysis concluded that stresses caused by thermal expansion of instrumentation cable/wire attachment tabs and the backside of the mirror led to the tabs separating from their attachment point, which removed small chips of glass from that portion of the mirror. The chips were repaired and the project has been cleared to continue flights. Delays associated with this issue resulted in the loss of three flight opportunities. As we reported previously, design challenges with the Cavity Door Drive System (CDDS) and longer than anticipated door testing, delayed initial science flights by one year. Project officials stated that the CDDS issues were being caused by a faulty sensor and have since improved the software. According to the project, the CDDS will fly as designed and SOFIA can successfully conduct an open door landing.

Other Issues to be Monitored: Due to developmental delays of SOFIA, the project continues to monitor the potential loss of science community support. According to the project, support for SOFIA has been increasing now that the achievable results—such as first light images obtained in May 2010—have become available.

PROJECT OFFICE COMMENTS

The SOFIA project provided technical comments to a draft of this assessment, which were incorporated as appropriate. Project officials also commented that the project completed early science and engineering flights for observatory characterization and the project met or exceeded 100 percent of all early science requirements. They added that the project accelerated the demonstration of an international deployment.

Tracking and Data Relay Satellite Replenishment

Recent/Continuing Project Challenges

- Contractor Issues
- Technology Issues
- Launch Issues

Previously Reported Challenges

- Parts Issues

Project Essentials

NASA Center Lead: **Goddard Space Flight Center**
Partner: **Non-NASA Agencies**
Projected Launch Date:
TDRS K – December 2012
TDRS L – December 2013
Launch Location: **Cape Canaveral AFS, FL**
Launch Vehicle: **Atlas V**
Mission Duration: **15 years**
Requirement derived from: **Support and expand existing TDRSS fleet**

Contract Information

Prime Contractor: **Boeing Satellite Systems**
Type of Contract: **Fixed Price Incentive Fee**
Date of Award: **December 2007**
Initial Value of Contract: **$1.38 billion***
Current Value: **$1.41 billion***

Source: © Boeing (artist depiction).

* This represents the full cost of the Boeing contract that NASA is managing; however, the cost is shared with NASA's partners.

The Tracking and Data Relay Satellite (TDRS) System consists of in-orbit communication satellites stationed at geosynchronous altitude coupled with two ground stations located in New Mexico and Guam. The satellite network and ground stations provide mission services for near-Earth user satellites and orbiting vehicles. TDRS K and L are the 11th and 12th satellites, respectively, to be built for the TDRS system. They will contribute to the existing network by providing continuous high bandwidth digital voice, video, and mission payload data, as well as health and safety data relay services to Earth-orbiting spacecraft such as the International Space Station and the Hubble Space Telescope.

Project Performance

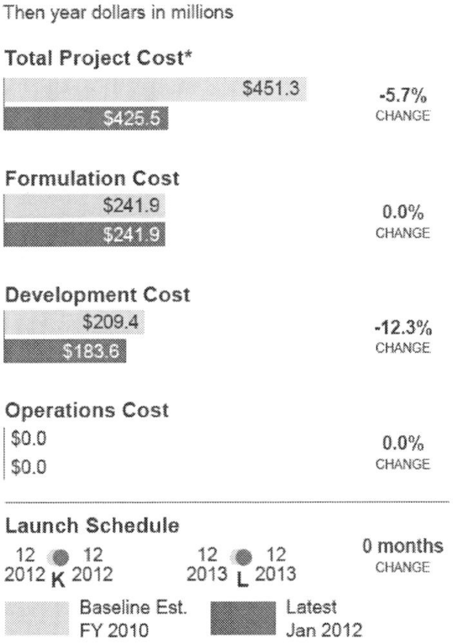

* Represents the estimate of NASA funding and does not include expected partner contributions.

Project Summary

The TDRS-K project was on the launch manifest for its internal management launch date of June 2012. However, contractor schedule performance and system level integration and test issues have depleted the schedule margin to this date and the project reported it was re-manifested to its December 2012 baseline launch date. Boeing is adding resources and pulling work forward on TDRS L in an effort to preserve its internal management launch date of February 2013; however, TDRS L is currently scheduled to launch in February 2014. Project officials report that the agency is actively working to find a launch slot in 2013 to meet its December 2013 commitment.

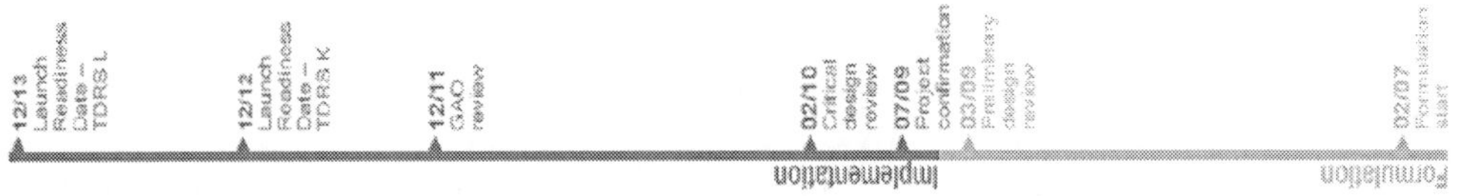

Project Update

Contractor Issues

The TDRS replenishment project has experienced cost overruns and depletions of schedule margin against the manifested June 2012 launch date that, according to project officials, was a result of subcontractor schedule performance and system level integration and test issues at the prime contractor—Boeing. According to project officials, Boeing has exceeded the ceiling price on the fixed price incentive fee contract and is responsible for costs exceeding the ceiling unless valid requests for equitable adjustments or claims are submitted to and accepted by NASA. Project officials stated that challenges faced by Boeing include late delivery of payload units from two of Boeing's subcontractors, and technical issues found during spacecraft bus and payload module integration and testing. For example, the Channel Control Unit exhibited a small non-compliance during testing that required further testing. This issue contributed to the payload module being delayed for system level thermal vacuum testing.

Technology Issues

Project officials reported delays due to merging new ground systems software code with the existing operational software baseline at the White Sands Complex. The project continues to track latent software defects as a risk because additional software defects could have cost and schedule impacts. In addition, project officials reported that heavy demands on the TDRS ground systems from existing satellites makes integration of TDRS K modifications a challenge. The project is waiting for completion of a study to determine if current resources will be adequate. However, project officials stated that if additional ground assets are needed, costs could increase as these assets can be very expensive.

Launch Issues

The TDRS K project continues to maintain its baseline launch readiness date of December 2012. Project officials stated that late payload unit deliveries and issues found during integration and testing, however, have depleted the schedule margin to its June 2012 internal management launch date. TDRS-K was on the launch manifest for June 2012; however, the project reports that the mission has been re-manifested for December 2012. Boeing is adding resources and pulling work forward on TDRS L in an effort to preserve its internal management launch date of February 2013, however, as a result of the manifest change for TDRS K, TDRS L is now manifested to launch in February 2014. Project officials report that the agency is actively working to find a launch slot after June 2013 to allow for adequate on-orbit testing on TDRS K before TDRS L is launched and to meet the December 2013 commitment.

Other Issues to be Monitored

NASA considers the TDRS system to be a national asset. The International Space Station is supported by, and many near-Earth spacecraft are dependent upon, the satellite system for communication. However, even with the successful launch of TDRS K and L, continuity of service for the TDRS System can only be ensured for NASA and other government agency users through approximately fiscal year 2016 at current support levels. The primary reason for this is due to an aging fleet of satellites currently in operations and the recent retirement of two spacecraft. According to project officials, the current fixed price incentive fee development contract with Boeing for TDRS K and L includes options to produce two additional TDRS satellites that could extend TDRS system service continuity. In November 2011, NASA elected to exercise the first option for TDRS-M at an estimated value of $289 million.

PROJECT OFFICE COMMENTS

The TDRS Replenishment project provided technical comments to a draft of this assessment, which were incorporated as appropriate. Project officials also commented that while the prime contractor has experienced subcontractor delivery and integration and test delays, the TDRS project is on track to satisfy the program commitment for a December 2012 launch of TDRS K. They added that the project has also made progress integrating and testing ground modifications in the operational environment at White Sands and the modifications are planned to be completed prior to the launch of TDRS K.

AGENCY COMMENTS AND OUR EVALUATION

We provided a draft of this report to NASA for review and comment. In its written response, NASA agreed with our findings and stated that it remains dedicated to continuous improvement of its acquisition management processes and performance and will continue to identify and address the challenges that lead to cost and schedule growth in its projects.

NASA commented on the information we presented on contract value changes. Specifically, NASA noted that there are multiple factors that could drive a change in a contract's value, such as evolving requirements and funding availability; however, NASA did not provide explanations of contract value changes for the majority of the projects we reviewed.

There are multiple factors that could lead to an increase or decrease in a contract's value, but we note that unexpected changes that significantly increase the contract's value could also increase the overall life-cycle cost of a project. Upon further reflection, we removed the information on contract value changes because we believe this is an area that warrants a more detailed examination and, as noted in our report, we plan to study the extent to which changes in contract value could impact cost and schedule baselines in the future. We look forward to working with NASA to ensure that it provides sufficient information to explain the reason(s) behind any changes to contract values.

NASA also stated that our definition of funding challenges is broader than its definition. We agree that the definition is broad in terms of scope. While there is value in defining funding challenges more consistently for individual projects, it is their cumulative effect that is of greater concern. For example, the agency is already feeling the effects of a constrained budget and ramifications of the significant cost increase experienced by the JWST project as evidenced by the termination of the EMTGO project.

As stated in the report, we view this challenge as an area that will require close attention in the next few years as NASA moves forward with JWST and even larger investments in its human spaceflight program while operating in a constrained budget environment.

Consistent with its comments on our previous reviews, NASA noted that its projects are high-risk, one-of-a-kind development efforts that do not lend themselves to all the practices of a "business case" approach that we outlined since essential attributes of NASA's project development differ from those of a production entity.

We recognize these differences and have made adjustments to our assessment methodology. For example, we do not assess NASA's science projects for production maturity and tailor our evaluation of technology maturity given testing constraints associated with its projects.

The basic premise, however, of developing a sound business case for a project does apply to NASA projects and helps ensure that the agency is making well-informed decisions based on high levels of knowledge about the risks associated with an individual project and how it fits into the broader portfolio of projects. NASA has recognized the importance of developing a sound business case and incorporated this concept into its program and project management policy document.

We continue to work with NASA officials on opportunities to refine the information they use to ensure adequate levels of knowledge exist at key points in a project life cycle and measure the progress of their projects. For example, given NASA's concerns with the drawing release metric we use, last year we recommended that NASA develop a common set of measurable and proven criteria to assess design stability and to amend NASA's systems engineering policy to that effect.

In response, NASA has developed a set of proposed indicators to monitor the progress of a project's design that it believes will provide a more comprehensive approach than it currently uses. NASA plans to update its policies and handbooks to include this information. We believe this approach has merit as the indicators will track the development of the design beginning at the preliminary design review and will bring more focused attention to the issue. NASA and GAO will need to monitor the use of these indicators over the life cycle of several projects to determine their effectiveness, as NASA currently does not have data to support such a determination.

We will continue to work with NASA to discuss the application of these and other metrics to asses design maturity; however, regardless of the metric used to monitor design progress, the need for a stable design to support sound and informed decision making is valid. Our work has shown that if project development continues without design stability, it is at a greater risk of costly re-designs to address changes to project requirements and unforeseen challenges.

We appreciate the work that NASA has done to begin to address this issue and look forward to the process of proving out these indicators.

NASA's written comments are reprinted in appendix I. NASA also provided technical comments, which we addressed and incorporated throughout the report as appropriate.

Cristina T. Chaplain
Director
Acquisition and Sourcing Management

APPENDIX I. COMMENTS FROM THE NATIONAL AERONAUTICS AND SPACE ADMINISTRATION

National Aeronautics and
Space Administration

Office of the Administrator
Washington, DC 20546-0001

February 16, 2012

Ms. Cristina Chaplain
Director
Acquisition and Sourcing Management
United States Government Accountability Office
Washington, DC 20548

Dear Ms. Chaplain:

The National Aeronautics and Space Administration (NASA) appreciates the opportunity to comment on the Government Accountability Office (GAO) draft report entitled "Assessments of Selected Large-Scale Projects" (GAO-12-207SP). NASA values the continued open and constructive communications between NASA and the GAO on this effort and appreciates the ongoing work by the GAO audit team. NASA remains dedicated to continuous improvement of its acquisition management processes and performance and will continue to work with the GAO to identify and address any challenges that may lead to cost and schedule growth of our projects.

NASA is pleased that the GAO recognized NASA's achievements in 2011 with the launch of 5 of the 15 projects evaluated as part of the GAO's assessment. Aquarius launched on June 10, 2011, and on September 22, 2011, the project released its first global map of the salinity of the ocean surface, providing an early glimpse of the mission's anticipated discoveries. Juno launched on August 5, 2011, and is now on its five-year journey to Jupiter. Gravity Recovery and Interior Laboratory (GRAIL) launched on September 10, 2011, and has already beamed back its first images of the far side of the moon. The instrument that acquired these images, MoonKAM, is dedicated to student-led efforts in selecting lunar images for study. NASA launched the National Polar-orbiting Operational Environmental Satellite System Preparatory Project (NPP) on October 28, 2011. The project was recently renamed Suomi National Polar-orbiting Partnership, or Suomi NPP, in honor of the late Verner E. Suomi, who is recognized widely as "the father of satellite meteorology." Lastly, the Mars Science Laboratory (MSL) launched on November 26, 2011, and is expected to arrive at Mars in August 2012.

NASA also recognizes that the continuing refinement, improvement, and implementation of sound acquisition practices, policies ,and processes are essential to reducing project cost and schedule growth on future similar missions. As highlighted in the draft report, NASA continues to implement initiatives to mitigate acquisition management risks. NASA instituted a Joint Cost and Schedule Confidence Level (JCL) policy in 2009 to increase likelihood of project success at the specified funding level. Implementation of the JCL process closely followed the adoption of a policy that requires projects to be budgeted at 70 percent confidence level for their estimated cost, as part of NASA's continuous improvement in cost estimating policies and processes.

The JCL is one of many inputs into the baseline decision process and has the desired effect of increasing insight by both the project manager and NASA's leadership by appropriately taking into account risk-based uncertainties associated with the integrated cost and schedule plan. As NASA continues to implement and improve the use of the JCL process, we will continue to assess the impact of this process on cost and schedule performance, as the projects baselined under this policy complete their development phase. To date, two projects that were subject to the 70 percent cost confidence level policy had successful launches (on schedule and within the estimated cost) – Juno and GRAIL. Additionally, the missions baselined are using the JCL process Magnetospheric Multiscale, Lunar Atmosphere and Dust Environment Explorer, and Mars Atmosphere and Volatile Evolution – currently remain on track to complete development within their cost baselines. NASA expects many of the current missions under development to be completed within cost and schedule as per the 70 percent JCL policy.

In the draft report, the GAO supplied a table that outlines the value change on multiple contracts for projects in implementation and noted plans to study contract value changes and the extent to which they could impact cost and schedule baselines. NASA looks forward to working with the GAO to facilitate a better understanding of the Agency's contracting policy and practices, with respect to management of contracts. The table on page 17: "Contract Value Changes for Selected Major NASA Projects" illustrates a change in contract value from initial award to current value. The difference between these two values could be attributed to many factors – most notably increases in content performed on the contract and do not necessarily represent unwarranted cost growth.

A change to contract value across the period of performance may result from planned, evolving requirements, funding availability at the time of initial award, funding changes that deviate from the planned profile, and/or the exercise of a contract option. For example, some of the contracts identified in the report were initially awarded in the project formulation phase and included an initial, narrow segment of work rather than the full spectrum of activities associated with a project's life cycle. In these cases, as projects matured from formulation into implementation, the scope of work was expanded to include requirements appropriate to the implementation phase.

The draft report identifies that 12 of the 21 projects assessed experienced funding challenges. While NASA agrees that projects have experienced funding challenges, the breadth of the GAO's definition appears broader than NASA's definition of this issue based on the projects assessed. Moving forward, NASA will work with the GAO to refine the definition of a funding challenge to strengthen the annual assessment and provide an improved construct for assessing projects on this issue.

As stated in previous years, NASA already implements many of the elements of the GAO's stated "knowledge-based business case" approach, where appropriate and applicable to a research and development project. While NASA business practices represent some alignment to this approach, there are essential attributes of NASA's project development that differ from those of a commercial and/or production entity. Specifically, NASA's spaceflight systems rarely have a production phase and, with few exceptions, are not based upon a spiral development process. NASA's missions extend the state of the art in the areas of science and technology by investigating new areas of space science and aeronautics. This is accomplished typically through the development of one-of-a-kind missions employing new or complex technologies and systems. As a result, these are high-risk activities.

NASA continues to work with the GAO to adapt its assessment methodology to better assess and reflect NASA's programs and projects. For example, as noted in the draft report, many of the projects in implementation experienced challenges with design stability, as indicated by the percentage of engineering drawings releasable at critical design review. NASA and the GAO have discussed an alternative measure to the drawing release metric, based on NASA's opinion that this is only one of many indicators to measure a project's readiness as they move forward from the critical design review. As shared with the GAO, NASA's plan is to enhance the ability to monitor and assess the stability of programs and projects, as well as the maturity of their products through the development process using a more comprehensive approach, including a set of proposed indicators as an alternative to the drawing release metric. This approach will be included in updates to our policies and handbooks for program and project management, as also recommended in House Report (House Report 112-69) accompanying the FY 2012 Commerce, Justice, Science, and related agencies appropriations bill.

Finally, NASA would like to thank the GAO for considering and incorporating many technical corrections provided by projects' subject matter experts as part of the audit. Inclusion of these comments is important to present an accurate and balanced view.

NASA is committed to working with the GAO jointly to address any questions. NASA appreciates the ongoing opportunity to comment on the assessments and will continue to work with the GAO to ensure that comments are submitted and appropriately reflected in final reports. If you have any questions or require additional information, please contact Tracy Osborne at (202) 358-3795.

Sincerely,

Lori B. Garver
Deputy Administrator

Appendix II. Objectives, Scope, and Methodology

Our objectives were to report on the status and challenges faced by NASA systems with life-cycle costs of $250 million or more and to discuss broader trends faced by the agency in its management of acquisitions. In conducting our work, we evaluated performance and identified challenges for each of 21 major projects. We summarized our assessments of each individual project in two components—a project profile and a detailed discussion of project challenges. We did not validate the data provided by the project offices, but reviewed the data and performed various checks to determine that the data were reliable enough for our purposes. Where we discovered discrepancies, we clarified the data accordingly. Where applicable, we confirmed the accuracy of NASA-generated data with multiple sources within NASA and, in some cases, with external sources.

We developed a standardized data collection instrument (DCI) that was completed by each project office. Through the DCI, we gathered basic information about projects as well as current and projected development activities for those projects. The cost and schedule data estimates that NASA provided were the most recent updates as of January 2012; NASA provided performance data as of May 2011 and updated this data for some projects through

January 2012. At the time we collected the data, 6 of the 21 projects were in the formulation phase and the remaining 15 projects were in the implementation phase. To further understand performance issues, we analyzed monthly status reviews for each project for which they were available and talked with officials from most of the project offices and NASA's Office of the Chief Financial Officer (OCFO), Strategic Investments Division (SID). We also collected cost and schedule data for projects in operations that we had reviewed in prior reports for historical purposes. These projects were Dawn, Gamma-ray Large Area SpaceTelescope, Glory, Herschel, Kepler, Lunar Reconnaisance Orbiter, Orbiting Carbon Observatory, Solar Dynamics Observatory, and Wide-field Infrared Survey Explorer.

The information collected from each project office, Mission Directorate, and OCFO/SID were summarized in a 2-page report format providing a project overview; key cost, contract, and schedule data; and a discussion of the challenges associated with the deviation from relevant indicators from best practice standards. The aggregate measures and averages calculated were analyzed for meaningful relationships, e.g., relationship between cost growth and schedule slippage and knowledge maturity attained both at critical milestones and through the various stages of the project life cycle. We identified cost and/or schedule growth as significant where, in either case, a project's cost and/or its schedule exceeded the thresholds that trigger reporting to the Congress.

To supplement our analysis, we relied on GAO's work over past years examining acquisition issues across multiple agencies. These reports cover such issues as contracting, program management, acquisition policy, and cost estimating. GAO also has an extensive body of work related to challenges NASA has faced with specific system acquisitions, financial management, and cost estimating. This work provided the context and basis for large parts of the general observations we made about the projects we reviewed. Additionally, the discussions with the individual NASA projects helped us identify further challenges faced by the projects. Together, the past work and additional discussions contributed to our development of a short list of challenges discussed for each project. The challenges we identified and discussed do not represent an exhaustive or exclusive list. They are subject to change and evolution as GAO continues this annual assessment in future years. The challenges, indicated as "issues," are based on our definitions and assessments, not that of NASA.

To assess NASA's efforts to improve its acquisition management, we requested Joint Cost and Schedule Confidence Levels (JCL) for the five projects that completed them. For each of the five projects, NASA provided us with a few slides that summarized each project's JCL analysis. These project JCLs were incomplete and received late in our review, affecting our ability to conduct a thorough analysis of the data. To determine whether NASA's was budgeting to the 70 percent confidence level established in the JCL policy, we compared the JCL cost estimate to NASA's Integrated Budget and Performance documents. We previously received independent cost estimates and Standing Review Board presentations on some of these projects. In most cases, we received independent cost estimates conducted at the center level by the projects, along with estimates by the Aerospace Corporation and/or by NASA's Independent Program Assessment Office. We also requested the Deputy Program Management Council's JCL decision memos, but these were not provided in time to include in our analysis. We interviewed NASA officials and officials from one of the contractors that helped to develop the JCL model to discuss the policy's development and implementation.

We also have ongoing work that is assessing whether NASA's large spaceflight projects are effectively using earned value management techniques to manage their acquisitions. The team performing this review requested earned value management data for 10 of the 21 projects in our review and plan to report its findings in the summer of 2012.

Our work was performed primarily at NASA headquarters in Washington, D.C. In addition, we visited NASA's Dryden Flight Research Center at Edwards Air Force Base in California; Goddard Space Flight Center in Greenbelt, Maryland; the Jet Propulsion Laboratory in Pasadena, California; Johnson Space Center in Houston, Texas; and Marshall Space Flight Center in Huntsville, Alabama, to discuss individual projects. We also met with officials from NASA's Ames Research Center at Moffett Field in California.

Data Limitations

NASA provided updated cost and schedule data as of January 2012 for projects in implementation, or 15 of the 21 projects in our review. NASA provided internal preliminary estimated total (life-cycle) cost ranges and associated schedules for three of the projects that had not yet entered implementation, which were established at key decision point B (KDP-B).[38] We did not receive cost estimates or ranges for three projects— ExoMars Trace Gas Orbiter, Orion Multi-Purpose Crew Vehicle, and Space Launch System—since these projects had not yet reached their KDP-B, the point in the acquisition life cycle where a preliminary life-cycle cost estimate would normally be developed. We did receive preliminary scheduled launch dates for one of these projects—ExoMars Trace Gas Orbiter. NASA formally establishes cost and schedule baselines, committing itself to cost and schedule targets for a project with a specific and aligned set of planned mission objectives, at key decision point C (KDP-C), which follows a preliminary design review (PDR). KDP-C reflects the life-cycle point where NASA approves a project to leave the formulation phase and enter into the implementation phase. NASA explained that preliminary estimates are generated for internal planning and fiscal year budgeting purposes at KDP-B, which occurs midstream in the formulation phase, and hence, are not considered a formal commitment by the agency on cost and schedule for the mission deliverables. NASA officials stated that because of changes that occur to a project's scope and technologies between KDP-B and KDP-C, estimates of project cost and schedule can change significantly heading toward KDP-C.

Project Profile Information on Each Individual 2-Page Assessment

This section of the 2-page assessment outlines the essentials of the project, its cost and schedule performance, and its summary. Project essentials reflect pertinent information about each project, including, where applicable, the major contractors and partners involved in the project. These organizations have primary responsibility over a major segment of the project or, in some cases, the entire project.

Project performance is depicted according to cost and schedule changes in the various stages of the project life cycle. To assess the cost and schedule changes of each project, we obtained data directly from NASA OCFO/SID and from NASA's Integrated Budget and Performance documents. We compared the current cost and schedule data reported by NASA in January 2012 to previously established project cost and schedule baselines to determine the extent to which each project exceeded its baselines.

All cost information is presented in nominal then-year dollars for consistency with budget data.[39] Baseline costs are adjusted to reflect the cost accounting structure in NASA's fiscal year 2009 budget estimates. For the fiscal year 2009 budget request, NASA changed its accounting practices from full-cost accounting to reporting only direct costs at the project level.

The schedule assessment is based on acquisition cycle time, which is defined as the number of months between the project's start, or formulation start, and projected or actual launch date.[40] Formulation start generally refers to the initiation of a project; NASA refers to a project's start as key decision point A, or the beginning of the formulation phase. The preliminary design review typically occurs toward the end of the formulation phase, followed by a review at key decision point C, known as project confirmation, which allows the project to move into the implementation phase.

The critical design review is generally held during the latter half of the final design and fabrication phase of implementation and demonstrates that the maturity of the design is appropriate to support continuing with the final design and fabrication phase. Launch readiness is determined through a launch readiness review that verifies that the launch system and spacecraft/payloads are ready for launch. The implementation phase includes the operations of the mission and concludes with project disposal.

Project Challenges Discussion on Each Individual 2-Page Assessment

To assess the project challenges for each project, we submitted a DCI to each project office. In the data collection instrument, we requested information on the maturity of critical and heritage technologies, number of releasable design drawings at project milestones, software development information, project contractors with related contract values and award fees, and project partnerships.

We also held interviews with representatives from each of the projects to discuss the information on the data collection instrument. These discussions led to identification of further challenges faced by NASA projects. The seven challenges we identified were largely apparent in the projects that had entered the implementation phase; however, there were instances where these challenges were identified in projects in the formulation phase. We then reviewed pertinent project documentation—such as project plans, schedules, risk assessments, and major project review documentation—to corroborate any testimonial evidence we received in the interviews.

To assess issues with technology, we asked project officials to provide the technology readiness levels of each of the project's critical technologies at various stages of project development. Originally developed by NASA, technology readiness levels are measured on a scale of one to nine, beginning with paper studies of a technology's feasibility and culminating with a technology fully integrated into a completed product. (See appendix IV for the definitions of technology readiness levels.)

In most cases, we did not validate the project offices' selection of critical technologies or the determination of the demonstrated level of maturity. However, we sought to clarify the technology readiness levels in those cases where the information provided raised concerns, such as where a critical technology was reported as immature late in the project development cycle. Additionally, we asked project officials to explain the environments in which technologies were tested.

Our best practices work has shown that a technology readiness level of 6—demonstrating a technology as a fully integrated prototype in a relevant environment—is the level of maturity needed to minimize risks for space systems entering product development. In our assessment, the technologies that have reached technology readiness level 6 are referred to as fully mature because of the difficulty of achieving technology readiness level 7, which is demonstrating maturity in an operational environment—space.

Projects with critical technologies that did not achieve maturity by the preliminary design review were assessed as having a technology issue project challenge. We did not assess technology maturity for those projects that had not yet reached the preliminary design review at the time of this assessment.[41]

We also asked project officials to assess the technology readiness level of each of the project's heritage technologies at various stages of project development and interviewed project officials about the use of heritage technologies in their projects. We asked them what heritage technologies were being used; what effort was needed to modify the form, fit, and function of the technology for use in the new system; whether the project encountered any problems in modifying the technology; and whether the project considered the heritage technology as a risk to the project. Heritage technologies were not considered critical technologies by several of the projects we reviewed. Based on our interviews, review of data from the data collection instruments, and previous GAO work on space systems, we determined whether these technology issues were a challenge for a particular project.

To assess design stability, we asked project officials to provide the percentage of engineering drawings completed or projected for completion by the preliminary and critical design reviews and as of our current assessment.[42]

In most cases, we did not verify or validate the percentage of engineering drawings provided by the project office. However, we collected the project offices' rationale for cases where it appeared that only a small number of drawings were completed by the time of the design reviews or where the project office reported significant growth in the number of drawings released after the critical design review.

In accordance with best practices, projects were assessed as having achieved design stability if they had at least 90 percent of projected drawings releasable by the critical design review. Projects that had not met this metric were determined to have a design stability project challenge. Though some projects used other methods to assess design stability, such as computer and engineering models and analyses, we did not assess the effectiveness of these other methods. We did not assess design stability for those projects that had not yet reached the critical design review at the time of this assessment.

To assess issues with launch, we interviewed NASA's Launch Services and project officials. Launch issues were considered a challenge if, after establishing a firm launch date, a project had difficulty rescheduling its launch date because the project was not ready; if the project could be affected by another project slipping its launch; or if there were launch vehicle fleet issues.

In addition, we assessed the status of launch vehicle selection for projects in formulation and considered it a challenge if the proposed timing for the launch vehicle's selection date falls after preliminary design review due to the availability of certified medium class launch vehicles.

To assess issues with contractor management, we interviewed project officials about their interaction and experience with contractors. We also interviewed contractor representatives from Northrop Grumman Aerospace Systems and Lockheed Martin Space Systems Company. They informed us about contractor performance problems pertaining to their workforce, the supplier base, and technical and corporate experience. We assessed a project as having this challenge if these contractor issues caused the project to experience a cost overrun, schedule delay, or decrease in mission capability. For projects that did not have a major contractor, we considered this challenge inapplicable to the project.

To assess issues with parts quality, we submitted a data collection instrument to all of the projects in the implementation phase that were scheduled to be operating in a space environment. In addition, we asked project officials to identify project components that encountered parts quality or availability problems during development. Additionally, we asked project officials to explain the environments in which the parts quality issues were discovered and any implication on the project's cost and schedule. We considered parts issues a challenge if there were actual or potential cost and/or schedule impacts to the project as a result of parts quality or availability, or if the project had to take special steps in order to address parts issues.

To assess issues with development partners, we interviewed NASA project officials about their interaction with international or domestic partners during project development. Development partner issues were considered a challenge for the project if project officials indicated that domestic or foreign partners were experiencing problems with project development that impacted the cost, schedule, or performance of the project for NASA. These challenges were specific to the partner organization or caused by a contractor to that partner organization. For projects that did not have an international or domestic development partner, we considered this challenge not applicable to the project.

To assess issues with funding, we interviewed officials from NASA's OCFO/ SID and NASA project officials, and also relied upon past interviews with project contractors about the stability of funding throughout the project life cycle. Funding issues were considered a challenge if officials indicated that a project's funding had been interrupted or delayed resulting in an impact to the cost, schedule, or performance of the project or if project officials indicated that the project's budget does not have sufficient funding in certain years based on the work expected to be accomplished.

We corroborated the funding changes and reasons with budget documents when available. The individual project offices were given an opportunity to comment on and provide technical clarifications to the 2-page assessments prior to their inclusion in the final product. We incorporated these comments as appropriate and where sufficient supporting documentation was provided.

We conducted this performance audit from March 2011 to March 2012 in accordance with generally accepted government auditing standards. Those standards require that we plan and perform the audit to obtain sufficient, appropriate evidence to provide a reasonable basis for our findings and conclusions based on our audit objectives. We believe that the evidence obtained provides a reasonable basis for our findings and conclusions based on our audit objectives.

APPENDIX III. SELECTED MAJOR NASA PROJECTS REVIEWED IN GAO'S ANNUAL ASSESSMENTS

We have reviewed 32 major NASA projects since our initial review in 2009. See table 3 below for a list of projects included in our assessments from 2009 to 2012 and whether the project was in formulation or implementation at the time of our review.

Table 3. Selected Major NASA Projects Reviewed in GAO's Annual Assessments

	2009	2010	2011	2012
Projects in formulation	Ares I	Ares I	Ares I***	EMTGO****
	GPM	GPM	ICESat-2	ICESat-2
	JWST	LDCM	Orion***	Orion MPCV
	LDCM	Orion	SMAP	SLS
	Orion		SPP	SMAP
				SPP
Projects in impleme-ntation	Aquarius	Aquarius	Aquarius	Aquarius*
	Dawn*	Glory	Glory**	GPM
	GLAST*	GRAIL	GPM	GRAIL*
	Glory	Herschel*	GRAIL	Juno*
	Herschel	Juno	Juno	JWST
	Kepler	JWST	JWST	LADEE
	LRO	Kepler*	LADEE	LDCM
	MSL	LRO*	LDCM	MAVEN
	NPP	MMS	MAVEN	MMS
	OCO**	MSL	MMS	MSL*
	SDO	NPP	MSL	NPP*
	SOFIA	RBSP	NPP	OCO-2
	WISE	SDO*	OCO-2	RBSP
		SOFIA	RBSP	SOFIA
		WISE*	SOFIA	TDRS Replenishment
			TDRS Replenishment	

Source: GAO Analysis of NASA data.
* NASA projects that have launched.
** NASA projects that have launched but failed to reach orbit.
*** NASA projects that were cancelled before entering implementation.
**** In February 2012, NASA proposed canceling the EMTGO project as part of its fiscal year 2013 budget request.

Appendix IV. Technology Readiness Levels

Technology readiness level	Description	Hardware	Demonstration environment
1. Basic principles observed and reported.	Lowest level of technology readiness. Scientific research begins to be translated into applied research and development. Examples might include paper studies of a technology's basic properties.	None (paper studies and analysis).	None.
2. Technology concept and/or application formulated.	Invention begins. Once basic principles are observed, practical applications can be invented. The application is speculative and there is no proof or detailed analysis to support the assumption. Examples are still limited to paper studies.	None (paper studies and analysis).	None.
3. Analytical and experimental critical function and/or characteristic proof of concept.	Active research and development is initiated. This includes analytical studies and laboratory studies to physically validate analytical predictions of separate elements of the technology. Examples include components that are not yet integrated or representative.	Analytical studies and demonstration of nonscale individual components (pieces of subsystem).	Lab.
4. Component and/or breadboard. Validation in laboratory environment.	Basic technological components are integrated to establish that the pieces will work together. This is relatively "low fidelity" compared to the eventual system. Examples include integration of ad-hoc hardware in a laboratory.	Low fidelity breadboard. Integration of nonscale components to show pieces will work together. Not fully functional or form or fit but representative of technically feasible approach suitable for flight articles.	Lab.
5. Component and/or breadboard validation in relevant environment.	Fidelity of breadboard technology increases significantly. The basic technological components are integrated with reasonably realistic supporting elements so that the technology can be tested in a simulated environment. Examples include high-fidelity laboratory integration of components.	High-fidelity breadboard. Functionally equivalent but not necessarily form and/or fit (size, weight, materials, etc). Should be approaching appropriate scale. May include integration of several components with reasonably realistic support elements/ subsystems to demonstrate functionality.	Lab demonstrating functionality but not form and fit. May include flight demonstrating breadboard in surrogate aircraft. Technology ready for detailed design studies.

Technology readiness level	Description	Hardware	Demonstration environment
6. System/ subsystem model or prototype demonstration in a relevant environment.	Representative model or prototype system, which is well beyond the breadboard tested for TRL 5, is tested in a relevant environment. Represents a major step up in a technology's demonstrated readiness. Examples include testing a prototype in a high fidelity laboratory environment or in simulated realistic environment.	Prototype. Should be very close to form, fit, and function. Probably includes the integration of many new components and realistic supporting elements/subsystems if needed to demonstrate full functionality of the subsystem.	High-fidelity lab demonstration or limited/restricted flight demonstration for a relevant environment. Integration of technology is well defined.
7. System prototype demonstration in a realistic environment.	Prototype near or at planned operational system. Represents a major step up from TRL 6, requiring the demonstration of an actual system prototype in a realistic environment, such as in an aircraft, vehicle, or space. Examples include testing the prototype in a test bed aircraft.	Prototype. Should be form, fit, and function integrated with other key supporting elements/subsystems to demonstrate full functionality of subsystem.	Flight demonstration in representative realistic environment such as flying test bed or demonstrator aircraft. Technology is well substantiated with test data.
8. Actual system completed and "flight qualified" through test and demonstration.	Technology has been proven to work in its final form and under expected conditions. In almost all cases, this TRL represents the end of true system development. Examples include developmental test and evaluation of the system in its intended weapon system to determine if it meets design specifications.	Flight qualified hardware	Developmental Test and Evaluation (DT&E) in the actual system application
9. Actual system "flight-proven" through successful mission operations.	Actual application of the technology in its final form and under mission conditions, such as those encountered in operational test and evaluation. In almost all cases, this is the end of the last "bug-fixing" aspects of true system development. Examples include using the system under operational mission conditions.	Actual system in final form	Operational Test and Evaluation (OT&E) in operational mission conditions

Source: GAO and its analysis of NASA data.

Appendix V. Significant Accomplishments of Projects That Have Launched

Ten major NASA projects that we have reviewed launched prior to October 2011 and have begun mission operations. Below is a description of each mission's objective, its launch date, and a list of some of its significant accomplishments.

Aquarius

Mission Objective: Investigate the links between the global water cycle, ocean circulation, and the climate—including measuring global sea surface salinity.
Launch Date: June 2011
Date Mission Operations Began: August 2011

Significant Accomplishments:

- **September 2011** – Released its first global map of ocean surface salinity

Dawn

Mission Objective: Make measurements of the two largest asteroids in our solar system, Vesta and Ceres.
Launch Date: September 2007
Date Mission Operations Began: October 2007

Significant Accomplishments:

- **July-August 2011** – Reached orbit around Vesta asteroid and completed survey science phase
- **September 2011** – Transitioned to high-altitude mapping orbit and began science operations at this orbit

Gamma-ray Large Area Space Telescope (GLAST)

Mission Objectives: (1) Understand the mechanisms of particle acceleration in astrophysical environments; (2) determine the high-energy behavior of gamma-ray bursts; (3) resolve and identify point sources with known objects; and (4) probe dark matter and the extra galactic background light in the early universe.
Launch Date: June 2008
Date Mission Operations Began: August 2008

Significant Accomplishments:

- **October 2008** – Identified the first pulsars shining only in gamma rays
- **February 2010** – Made the most precise measurement of the high-energy cosmic-ray electron spectrum
- **March 2010** – Identified best astrophysical limits on some types of dark matter
- **September 2011** – Released its second catalog of objects, producing an inventory of 1,873 objects shining with the highest-energy form of light

Gravity Recovery and Interior laboratory (GRAIL)

Mission Objectives: Determine the structure of the lunar interior from crust to core, advance our understanding of the thermal evolution of the Moon, and extend our knowledge gained from the Moon to other terrestrial-type planets.

Launch Date: September 2011

Date Mission Operations Began: October 2011

Significant Accomplishments:

- **September 2011** – Completed pre-lunar orbit insertion spacecraft and payload checkout activities
- **October 2011** – Started operations
- **December 2011/January 2012** – Entered lunar orbit

Herschel

Mission Objectives: Seek to discover how the first galaxies formed and how they evolved to give rise to present day galaxies like our own.

Launch Date: May 2009

Date Mission Operations Began: November 2009

Significant Accomplishments:

- **April 2011** – Revealed an intricate network of filamentary structure, with new stars forming at locations where filaments intersect each other
- **August 2011** – Identified new molecules in interstellar clouds, including molecular oxygen
- **September 2011** – Detected water in the ring system of Saturn, which is likely coming from “volcanoes” on Saturn’s moon, Enceladus
- **October 2011** – Demonstrated that the heavy water to normal water ratio in the Hartley 2 Kuiper Belt comet is identical to that found in Earth’s oceans

Juno

Mission Objectives: Improve our understanding of the origin and evolution of Jupiter.
Launch Date: August 2011
Date Mission Operations Began: September 2011

Significant Accomplishments:

- **August 2011** – Deployed solar panel
- **August 2011** – Conducted low voltage instrument check out
- **September 2011** – Started operations
- **December 2011** – Scheduled to complete high voltage instrument check out

Kepler

Mission Objective: Discover Earth-like planets in orbit around stars in our galaxy.
Launch Date: March 2009
Date Mission Operations Began: May 2009

Significant Accomplishments:

- **January 2011** – Discovered first rocky planet outside our solar system
- **February 2011** – Identified 1,235 planet candidates since mission began
- **February 2011** – Discovered the most compact planetary system
- **September 2011** – Identified the first planet to orbit two stars

Lunar Reconnaissance Orbiter (LRO)

Mission Objective: Orbit the moon for one year measuring lunar topography, resources, temperatures, and radiation.
Launch Date: June 2009
Date Mission Operations Began: September 2009

Significant Accomplishments:

- **August 2010** – Discovered evidence of geologically recent tectonic activity on the Moon
- **September 2010** – Made 4-billion precise measurements of lunar topography, developing highly accurate maps for lunar science and exploration
- **October 2010** – Discovered the coldest places in the solar system are at the lunar pole
- **July 2011** – Identified a far-side lunar volcano

Solar Dynamics Observatory (SDO)

> **Mission Objective:** Understand the solar variations that influence life on Earth and humanity's technological systems.
> **Launch Date:** February 2010
> **Date Mission Operations Began:** May 2010

Significant Accomplishments:

- **June 2011** – Spotted the iconic surfer's wave rolling through the atmosphere of the sun, providing insight into how energy moves through the corona
- **August 2011** – Detected nascent sunspots deep below the solar surface before they are visible to the human eye
- **September 2011** – Discovered some solar flares have a "late phase flare" some minutes to hours later that has never before been fully observed and that pumps more energy into space than previously realized

Wide-field Infrared Survey Explorer (WISE)

> **Mission Objective:** Designed to map the sky in infrared light and search for the nearest and coolest stars, the origins of stellar and planetary systems, the most luminous galaxies in the universe, and most main-belt asteroids larger than 3 kilometers.
> **Launch Date:** December 2009
> **Date Mission Operations Began:** January 2010

Significant Accomplishments:

- **July 2010** – Completed survey on the entire sky
- **July 2011** – Discovered first Earth Trojan asteroid—those that share an orbit with a planet near stable points in front of or behind the planet
- **August 2011** – Identified the coldest class of stars, known as Y dwarfs
- **September 2011** – Found more than 90 percent of the largest near-Earth asteroids (greater than 1 km)

End Notes

[1] In June 2011, NASA began the process for closing out the Constellation program, which included the Ares I and Orion projects. This process included identifying those Constellation program elements that would be transitioned for use on the new Space Launch System and Orion Multi-Purpose Crew Vehicle programs.

[2] See Explanatory Statement, 155 Cong. Rec. H1653, 1824-25 (daily ed., Feb. 23, 2009), to the Omnibus Appropriations Act, 2009, Pub. L. No. 111-8. In this report, we refer to these projects as major projects rather than large-scale projects as this is the term used by NASA.

[3] NASA is required to report to Congress if the development cost of a program is likely to exceed the baseline estimate by 15 percent or more, or if a milestone is likely to be delayed by 6 months or more. 51 U.S.C. § 30104(e).

[4] GAO, *Best Practices: Using a Knowledge-Based Approach to Improve Weapon Acquisition*, GAO-04-386SP (Washington, D.C.: January 2004).

[5] GAO, *Defense Acquisitions: Key Decisions to Be Made on Future Combat System*, GAO-07-376 (Washington, D.C.: Mar. 15, 2007); *Defense Acquisitions: Improved Business Case Key for Future Combat System's Success*, GAO-06-564T (Washington, D.C.: Apr. 4, 2006); *NASA: Implementing a Knowledge-Based Acquisition Framework Could Lead to Better Investment Decisions and Project Outcomes*, GAO-06-218 (Washington, D.C.: Dec. 21, 2005); and *NASA's Space Vision: Business Case for Prometheus 1 Needed to Ensure Requirements Match Available Resources*, GAO-05-242 (Washington, D.C.: Feb. 28, 2005).

[6] GAO .

[7] NASA defines the formulation phase as the identification of how the program or project supports the agency's strategic goals; the assessment of feasibility, technology, concepts, and performance of trade studies; risk assessment and possible risk mitigations and continuous risk management processes; team building, development of operations concepts and acquisition strategies; establishment of high-level requirements and success criteria; the preparation of plans, budgets, and schedules essential to the success of a program or project; and the establishment of control systems to ensure performance to those plans and alignment with current agency strategies. *NASA Interim Directive (NID) NM 7120-97 for NASA Procedural Requirements (NPR) 7120.5D*, paragraph 1.5.1(a) (Sept. 28, 2011) (Hereinafter cited as NID for NPR 7120.5D (Sept. 28, 2011)).

[8] The implementation phase is defined as the execution of approved plans for the development and operation of the program or project, and the use of control systems to ensure performance to approved plans and requirements and continued alignment with the agency's strategic goals. NID for NPR 7120.5D, paragraph 1.5.1(c) (Sept. 28, 2011).

[9] According to NID for NPR 7120.5D, Table 2-4 (Sept. 28, 2011), the PDR evaluates the completeness/consistency of the planning, technical, and cost/schedule baselines developed during formulation. It assesses compliance of the preliminary design with applicable requirements, and determines if the project is sufficiently mature to begin the final design and fabrication phase.

[10] The agency baseline commitment is the integrated set of requirements, cost, schedule, technical content, and an agreed-to joint confidence level that forms the basis for NASA's commitment with OMB and Congress. NID for NPR 7120.5D, Appendix A (Sept. 28, 2011).

[11] According to NID for NPR 7120.5D, Table 2-4 (Sept. 28, 2011), the CDR evaluates the integrity of the project design and its ability to meet mission requirements, with appropriate margins and acceptable risk, within defined project constraints, including available resources. It determines if the design is appropriately mature to continue with the final design and fabrication phase.

[12] The system integration review (SIR) evaluates the readiness of the project and associated supporting infrastructure to begin system assembly, integration, and test. SIR evaluates whether the remaining project development can be completed within available resources and determines if the project is sufficiently mature to begin phase D, where test and integration activities occur. NID for NPR 7120.5D, Table 2-4 (Sept. 28, 2011).

[13] The life-cycle cost estimate for LADEE does not include $65.3 million for the Lunar Laser Communications Demonstration being developed by the Human Exploration and Operations Mission Directorate.

[14] GAO, *High-Risk Series: An Update*, GAO-11-278 , (Washington, D.C.: February 2011).

[15] GAO, *NASA: Implementing a Knowledge-Based Acquisition Framework Could Lead to Better Investment Decisions and Project Outcomes*, GAO-06-218 (Washington, D.C.: Dec. 21, 2005).

[16] GAO, *Additional Cost Transparency and Design Criteria Needed for National Aeronautics and Space Administration (NASA) Projects*, GAO-11-364R (Washington, D.C.: Mar. 3, 2011).

[17] We based our cost and schedule analysis on the 15 projects that were in the implementation phase of the project life cycle during the course of our review. NASA did not provide formal cost and schedule baselines for the six projects in formulation, citing that the estimates are preliminary. Baselines are established when the project transitions to implementation.

[18] For purposes of our analysis, cost or schedule growth is significant if it exceeds the thresholds that trigger reporting to Congress under the law. The thresholds are development cost growth of 15 percent or more from the baseline cost estimate or a milestone delay of 6 months or more beyond the baseline schedule estimate. 51 U.S.C. § 30104(e).

[19] For this analysis, we examined each project's largest contract with a value of $20 million or more at award. The Aquarius, LADEE, and MSL projects were not included because they are in-house development projects either at a NASA center or at the Jet Propulsion Laboratory—a federally funded research and development center.

[20] Appendix IV provides a description of the metrics used to assess technology maturity.

[21] NASA Procedural Requirements 7123.1A, NASA *Systems Engineering Processes and Requirements,* Appendix G, paragraph G.19(b) (Mar. 26, 2007).

[22] One project that did not meet the criteria was no longer included in our review—Glory—and another project—SMAP—held its preliminary design review, but did not meet the criteria.

[23] We were unable to assess design stability for the SOFIA project as NASA has lost this data. According to project officials, the project documentation did not transfer in its entirety from Ames Research Center to Dryden Flight Research Center.

[24] Engineering drawings are considered to be a good measure of the demonstrated stability of a product's design because the drawings represent the language used by engineers to communicate to the manufacturers the details of a new product design—what it looks like, how its components interface, how it functions, how to build it, and what critical materials and processes are required to fabricate and test it. Once the design of a product is finalized, the drawing is "releasable."

[25] GAO ; GAO, *NASA: Issues Implementing the NASA Authorization Act of 2010*, GAO-11-216T (Washington, D.C.: Dec. 1, 2010); GAO-06-218 .

[26] A request for action is a formal written request sponsored by the review panel asking for additional information or action by the project team. It is generally developed as a result of insufficient safety, technical, or programmatic information being available at the time of the review.

[27] NASA's approach to mitigating a launch vehicle's risk is through development of a launch vehicle certification process, which is laid out in NASA Policy Directive 8610.7D, *Launch Services Risk Mitigation Policy for NASA-Owned and/or NASA-Sponsored Payloads/ Missions* (Jan. 31, 2008).

[28] While many Delta II launch vehicle components are no longer in production, NASA decided to include the five remaining Delta II launch vehicles on its launch services contract in September 2011. The Delta II launch vehicle had not been included on NASA's Launch Services II contract when it was awarded in September 2010.

[29] See 51 U.S.C. § 50134.

[30] GAO, *Space and Missile Defense Acquisitions: Periodic Assessment Needed to Correct Parts Quality Problems in Major Programs*, GAO-11-404 (Washington, D.C.: Jun. 24, 2011).

[31] GAO .

[32] NASA Policy Directive 1000.5A, *Policy for NASA Acquisition*, paragraph 1(h)(3) (Jan. 15, 2009).

[33] This is the point on the joint cost and schedule probability distribution where there is a 70 percent probability that the project will be completed at or lower than the estimated amount and at or before the projected schedule. NASA Policy Directive 1000.5A, *Policy for NASA Acquisition*, paragraph 1(h)(1)(a) (Jan. 15, 2009).

[34] This funding refers to the amount provided to the project office in order to execute the project. NASA Policy Directive 1000.5A, *Policy for NASA Acquisition*, paragraph 1(h)(2) (Jan. 15, 2009).

[35] GPM also prepared a JCL, but NASA officials told us the project has since gone through a replan and will complete another JCL to support the fiscal year 2013 budget submission.

[36] American National Standards Institute/Electronic Industries Alliance Standard, Earned Value Management Systems, ANSI/EIA-748-B-2007 approved July 9, 2007.

[37] GAO, *Cost Estimating and Assessment Guide: Best Practices for Developing and Managing Capital Program Costs*, GAO-09-3SP (Washington, D.C.: Mar. 2009).

[38] These missions include Ice, Cloud, and Land Elevation Satellite-2 (ICESat-2), Soil Moisture Active and Passive, and Solar Probe Plus.

[39] Because of changes in NASA's accounting structure, its historical cost data are relatively inconsistent. As such, we used then-year dollars to report data consistent with the data NASA reported to us.

[40] Some projects reported that their spacecraft would be ready for launch sooner than the date that the launch authority could provide actual launch services. In these cases, we used the actual launch date for our analysis rather than the date that the project reported readiness.

[41] According to NASA officials, projects that were in formulation at the time of the agency's 2007 revision of its project management policy are required to comply with that policy. Projects that had already entered implementation at the time of the revision were directed to implement those requirements that would not adversely affect the project's cost and schedule baselines.

[42] In our calculation for the percentage of total number of drawings projected for release, we used the number of drawings released at the critical design review as a fraction of the total number of drawings projected, including where a growth in drawings occurred. So, the denominator in the calculation may have been larger than what was projected at the critical design review. We believe that this more accurately reflected the design stability of the project.

In: Major Projects at NASA: Select Assessments
Editor: Eloy D'Aramitz ISBN: 978-1-62417-866-5

Chapter 2

JAMES WEBB SPACE TELESCOPE: ACTIONS NEEDED TO IMPROVE COST ESTIMATE AND OVERSIGHT OF TEST AND INTEGRATION*

United States Government Accountability Office

WHY GAO DID THIS STUDY

JWST is one of NASA's most expensive and technologically advanced science projects, intended to advance understanding of the origin of the universe. In 2011, JWST was rebaselined with a life cycle cost estimate of $8.8 billion and a launch readiness date in October 2018— almost nine times the cost and more than a decade later than originally projected in 1999. Concern about the magnitude of JWST's cost increase and schedule delay and their effects on NASA's progress on other high-priority missions led conferees for the Consolidated and Further Continuing Appropriations Act, 2012, to direct GAO to report on the project. Specifically, GAO assessed (1) the extent to which NASA's revised cost and schedule estimates are reliable based on best practices, (2) the major risks and technological challenges JWST faces, and (3) the extent to which NASA has improved oversight of JWST. To do this, GAO compared NASA's revised cost and schedule estimates with best practice criteria, reviewed relevant contractor and NASA documents, and interviewed project and contractor officials.

WHAT GAO RECOMMENDS

GAO recommends NASA take six actions including, among others, to take steps to improve its cost estimate; to conduct an additional, earlier independent review of test and integration activities; and to develop a long-term oversight plan that anticipates planned travel budget reductions. In commenting on a draft of this report, NASA fully or partially concurred

* This is an edited, reformatted and augmented version of United States Government Accountability Office, Publication No. GAO-13-4, dated December 2012.

with the recommendations citing ongoing efforts, but GAO believes some do not go far enough.

What GAO Found

The National Aeronautics and Space Administration (NASA) has provided significantly more time and money to the James Webb Space Telescope (JWST) than previously planned and expressed high confidence in the project's new baselines. Its current cost estimate reflects some features of best practices for developing reliable and credible estimates. For example, the estimate substantially meets one of four cost characteristics—comprehensive—that GAO looks for in a reliable cost estimate, in part because all life cycle costs were included. The estimate, however, only partially met the other three characteristics—well documented, accurate, and credible—which detracts from its reliability. For example, the estimate's accuracy, and therefore the confidence level assigned to the estimate, was lessened by the summary schedule used for the joint cost and schedule risk analysis because it did not provide enough detail to determine how risks were applied to critical project activities. The estimate's credibility was also lessened because officials did not perform a sensitivity analysis that would have identified key drivers of costs, such as workforce size. Program officials believe that it would have been difficult to fully address all best practice characteristics. GAO believes there is time to improve the estimate and enhance the prospects for delivering the project according to plan.

Project officials report that the JWST schedule has 14 months of reserve, which meets Goddard guidance for schedule reserve; however, only 7 of the 14 months are likely to be available for the last three of JWST's five complex integration and test efforts. GAO's prior work shows that the integration and test phases are where problems are commonly found and schedules tend to slip. Given that JWST has a challenging integration and test schedule, this could particularly be likely. The project has made some significant progress in the past year, notably successfully completing development of the 18 primary mirror segments—considered JWST's top technical risk. Nevertheless, ongoing challenges are indicative of the kinds of issues that can require significant effort to address. For example, instrument challenges have delayed the first integration and test effort. In addition, key long-term risks on subsystems with a significant amount of work remaining will not be retired until 2016. Currently, NASA's plan for project oversight calls for one independent mission-level system integration review about 13 months before launch. While this is consistent with what NASA requires for its projects, this approach may not be sufficient for a project as complex as JWST.

JWST has taken several steps to improve communications and oversight of the project and its contractors—such as taking over responsibility for mission systems engineering from the prime contractor; instituting meetings that include various levels of NASA, contractor, and subcontractor management; and implementing a new risk management system to allow for better tracking of risks. The enhancements to the oversight of the project are steps in the right direction, but it will take time to assess their effectiveness and ensure that the efforts are sustained by the project in the future. Reductions in travel budgets, however, could require the project to adjust the oversight approach that was adopted as a result of the replan.

Additional reductions in travel budgets are anticipated in future years, but officials do not have a plan to address such reductions and their potential impact on continuing the current oversight approach.

ABBREVIATIONS

CSA	Canadian Space Agency
ESA	European Space Agency
FGS	Fine Guidance Sensor
ICRP	Independent Comprehensive Review Panel
ISIM	Integrated Science Instrument Module
JCL	Joint Cost and Schedule Confidence Level
JWST	James Webb Space Telescope
KDP	key decision point
MIRI	Mid-Infrared Instrument
NASA	National Aeronautics and Space Administration
NIRCam	Near-Infrared Camera
NIRISS	Near-Infrared Imager and Slitless Spectrograph
NIRSpec	Near-Infrared Spectrograph
NPR	NASA Procedural Requirements
OTE	Optical Telescope Element
OTIS	Optical Telescope Element and Integrated Science Instrument Module

December 3, 2012
Congressional Committees

The James Webb Space Telescope (JWST) is one of the National Aeronautics and Space Administration's (NASA) most expensive and technologically advanced science projects. It is intended to revolutionize our understanding of star and planet formation and advance the search for the origins of our universe. The project has been hampered by ineffective management, unrealistic estimates of costs, and insufficient resources, and is projected to cost significantly more and to take much longer to complete than expected when it was authorized to proceed in 1999. In 2011, JWST was rebaselined with a lifecycle cost estimate of $8.8 billion and a launch date in October 2018, which is almost nine times the cost and more than a decade later than originally projected. The magnitude of JWST's cost increase and schedule delay has and will continue to have a major effect on NASA's ability to implement or begin high-priority science missions for years to come.

NASA has expressed high confidence in being able to execute the project within its new cost and schedule baselines, and officials report that the project is being executed to its revised estimates in the year since the rebaseline. The project, however, is less than halfway through its development with known and unknown technical risks and challenges remaining to be addressed. The on-time and on-budget delivery of JWST is a high congressional priority. In November 2011, the Joint Explanatory Statement of the Committee of Conference for the Consolidated and Further Continuing Appropriations Act, 2012, recommended that JWST's formulation and development costs, not including the cost to operate the mission after launch, be no more than $8 billion. The conferees also directed GAO to assess the program annually and to report to the Committees on Appropriations on key issues relating to program and risk management, achievement of cost and schedule goals, and program technical status. This report responds to that direction. Specifically, we assessed (1) the extent to which NASA's revised cost and schedule estimates are reliable based on GAO best practices, (2) the major risks and technological challenges the JWST project faces, and (3) the extent to which NASA has improved the oversight of the JWST project.

Our approach included an examination of the revised cost and schedule estimates NASA developed during the replanning process using GAO's best practice criteria.[1] This included review of documentation of the inputs that supported the detailed cost and schedule estimates used by decision-makers to rebaseline the program and interviews with project officials involved in developing the estimates. We also reviewed the proposal from the JWST prime contractor used as input to the replan and interviewed contractor officials. We did not assess the most current JWST integrated master schedule because of ongoing negotiations for contract modifications associated with the replan. To identify the major short and long term risks and technological challenges facing the project, we interviewed project officials for the major subsystems as well as contractor officials. We also reviewed the project's risk database, monthly status reviews, and other documentation provided by the project officials. To understand the steps NASA implemented to improve the oversight being performed on the project, we interviewed project and contractor officials and reviewed documentation from independent reviews prior to the replan and actions taken by NASA in response. We also interviewed the Director of NASA's Office of Evaluation concerning a recent independent review of the JWST project performed by his office. We took appropriate steps to address data reliability, such as clarifying data discrepancies and corroborating NASA-generated data with other sources where applicable. NASA provided comments and technical clarifications on a draft of this report, which were incorporated as appropriate. Appendix I contains detailed information on our scope and methodology.

We conducted this performance audit from February 2012 to December 2012 in accordance with generally accepted government auditing standards. Those standards require that we plan and perform the audit to obtain sufficient, appropriate evidence to provide a reasonable basis for our findings and conclusions based on our audit objectives. We believe that the evidence obtained provides a reasonable basis for our findings and conclusions based on our audit objectives.

BACKGROUND

Initially referred to as the "Next Generation Space Telescope," JWST is a large deployable, infrared-optimized space telescope intended to be the successor to the aging Hubble Space Telescope. JWST is designed to be a 5-year mission to find the first stars and trace the evolution of galaxies from their beginning to their current formation, and is intended to operate in an orbit approximately 1.5 million kilometers—or 1 million miles—from the Earth. In a 2001 decadal survey, the National Research Council rated the JWST as the top-priority new initiative for astronomy and physics. With its 6.5-meter primary mirror, JWST will be able to operate at 100 times the sensitivity of the Hubble Space Telescope. A tennis-court-sized sunshield will protect the mirrors and instruments from the sun's heat to allow the JWST to look at very faint infrared sources. The Hubble Space Telescope operates primarily in the visible and ultraviolet regions.[2]

History of Cost Growth and Schedule Delays

JWST has experienced significant increases to project costs and schedule delays. Prior to being approved for development, cost estimates of the project ranged from $1 billion to $3.5 billion with expected launch dates ranging from 2007 to 2011. In March 2005, NASA increased the JWST's life-cycle cost estimate to $4.5 billion and slipped the launch date to 2013. We reported in 2006 that about half of the cost growth was due to schedule slippage—a 1-year schedule slip because of a delay in the decision to use a European Space Agency-supplied Ariane 5 launch vehicle and an additional 10-month slip caused by budget profile limitations in fiscal years 2006 and 2007. More than a third of the cost increase was caused by requirements and other changes. An increase in the program's contingency funding accounted for the remainder—about 12 percent—of the growth. NASA Headquarters chartered an Independent Review Team to evaluate the project that same year. In April 2006, the review team's assessment confirmed that the program's technical content was complete and sound, but expressed concern over the project's contingency reserve funding—funding used to mitigate issues that arise but which were previously unknown—reporting that it was too low and phased in too late in the development life cycle. The team reported that for a project as complex as the JWST, a 25 to 30 percent total contingency was appropriate. At that time, JWST's total contingency was about 19 percent. The team cautioned that this contingency compromised the project's ability to resolve issues, address risk areas, and accommodate unknown problems. The team also concluded that the 2013 launch date was not viable for the project based on its anticipated budget. It recommended that before the project was formally approved for development and baselined, NASA should take steps to provide the JWST project with adequate time-phased reserve funding to secure a stable launch date. Additional reserves were added and the project was baselined in April 2009 with a life-cycle cost estimate of $4.964 billion and a launch date in June 2014. Shortly after JWST was approved for development and its cost and schedule estimates were baselined, project costs continued to increase. In 2010, Senator Barbara Mikulski, chair of the Senate Committee on Appropriations, Subcommittee on Commerce, Justice, Science, and Related Agencies, asked NASA to initiate another independent review in response to the project's cost increases and reports that the June 2014 launch date was in jeopardy. The Independent Comprehensive Review Panel (ICRP) was commissioned by NASA and began its review in August 2010. In October 2010, the ICRP issued its report and cited several reasons for the project's problems including management, budgeting, oversight, governance and accountability, and communication issues.[3] The panel concluded JWST was executing well from a technical standpoint, but that the baseline funding did not reflect the most probable cost with adequate reserves in each year of project execution, resulting in an unexecutable project. The review panel recommended that additional resources be considered along with organizational and management restructuring. Following this review, the JWST program underwent a replan in

2011. In November 2011, the JWST project was reauthorized, but not before it was recommended for termination by the House Appropriations Committee. On the basis of the replan, NASA announced that the project would be rebaselined at $8.835 billion—a 78 percent increase to the project's life-cycle cost from the confirmed baseline—and would launch in October 2018—a delay of 52 months. The revised life-cycle cost estimate included 13 months of funded schedule reserve. In the President's Fiscal Year 2013 budget request, NASA reported a 66 percent joint cost and schedule confidence level associated with these estimates. A joint cost and schedule confidence level (JCL) is the process NASA uses to assign a percentage to the probable success of meeting cost and schedule targets and is part of the project's estimating process.

Current JWST Organization

The JWST project is divided into three major segments: the launch segment, the ground segment, and the observatory segment. The launch segment is primarily provided by the European Space Agency (ESA), which is contributing the Ariane 5 launch vehicle and launch site operations in French Guiana.

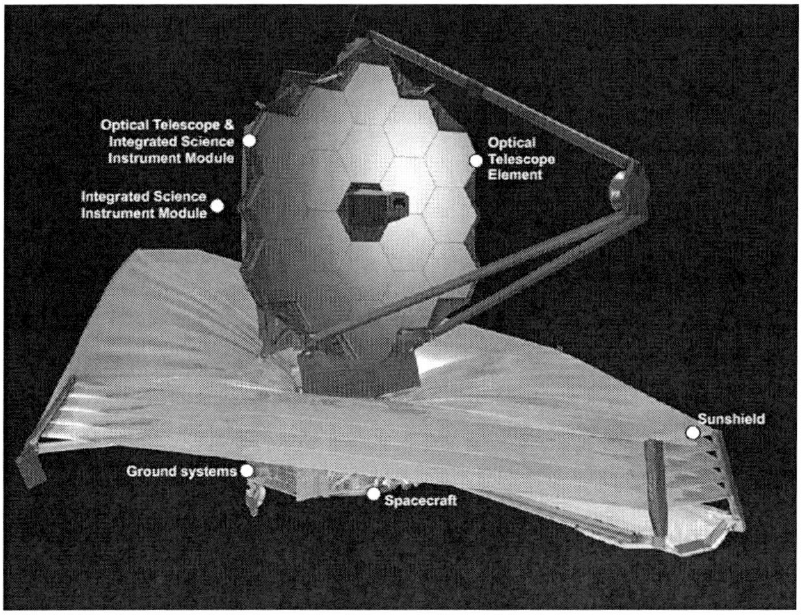

Sources: GAO (analysis); NASA (data and images).

Figure 1. James Webb Space Telescope. See Appendix II For Additional Information.

The ground segment will be responsible for collecting the data obtained by JWST in space and making it usable for scientists and researchers. This includes the development of software that will translate data into usable formats as well as operation of the software once the telescope is in space. The Space Telescope Science Institute, operated by the Association of Universities for Research in Astronomy (AURA) on a contract awarded by NASA, which currently performs science operations for the Hubble Space Telescope, is developing the science and operations and flight operations center for JWST and will conduct the first 6 months of flight and science operations. The NASA contract with the Space Telescope Science Institute extends through the first 6 months of JWST operations. A contract to manage the long term operations of JWST is planned to be awarded approximately 2 years prior to launch. The observatory segment will be launched into space and includes five major subsystems.[4] These subsystems are being developed through a mixture of NASA, contractor, and international partner efforts. See figure 1.

JWST is a single project program reporting directly to the NASA Associate Administrator for programmatic oversight and to the Associate Administrator for the Science Mission Directorate for technical and analysis support.[5] Goddard Space Flight Center is the NASA center responsible for the management of JWST. See figure 2 for the current JWST organizational chart.

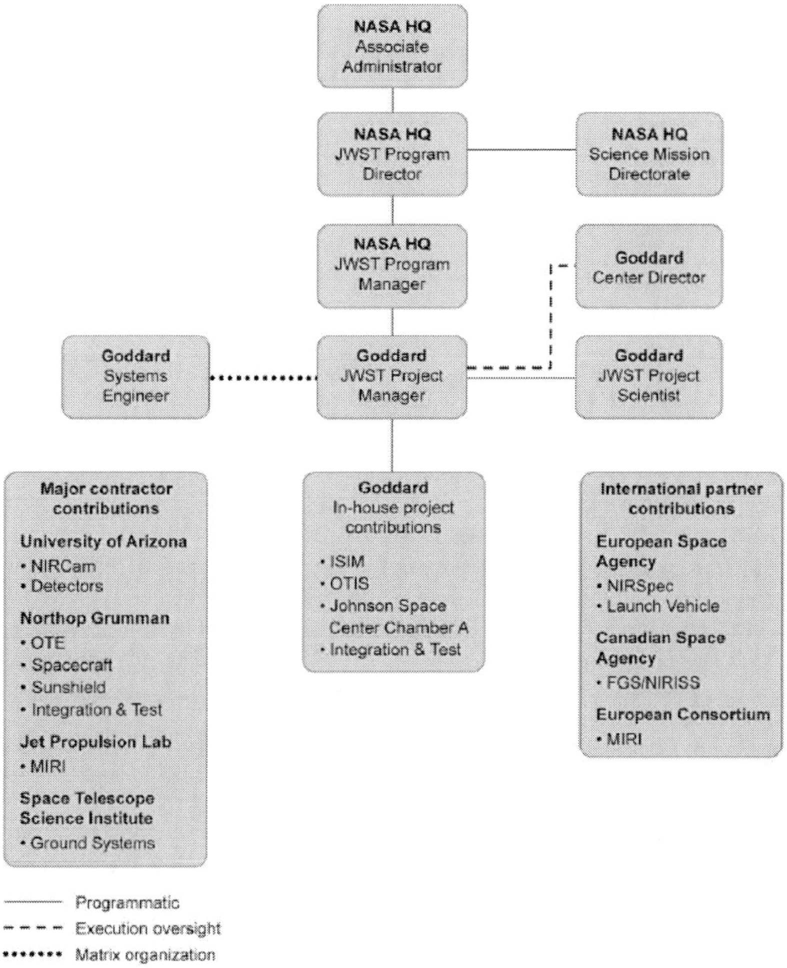

Source: GAO analysis of NASA information.

Figure 2. Organizational Chart for the JWST Program.

REVISED JWST COST ESTIMATE IS NOT FULLY CONSISTENT WITH BEST PRACTICES FOR RELIABLE AND CREDIBLE ESTIMATES AND THE INTEGRATED MASTER SCHEDULE IS NOT FINALIZED

Our analysis of JWST's revised cost estimate showed that it is not fully consistent with best practices for developing reliable and credible estimates, although project officials took some steps in line with best practices in the development of the estimate.

Characteristics	Best Practices Criteria	Overall assessment
Comprehensive	Include costs of the program over its full life-cycle, provide a level of detail appropriate to ensure that cost elements are neither omitted nor double-counted, and document all cost-influencing ground rules and assumptions	Substantially met
Well documented	Be supported by detailed documentation describing: • the purpose of the estimate, • the program background and system description, • the scope of the estimate, • the ground rules and assumptions, • all data sources, • estimating methodology and rationale, and • the results of the risk analysis. This documentation should show how the data used to derive the estimate can be traced back to, and verified against, their sources.	Partially met
Accurate	Be based on an assessment of most likely costs (adjusted for inflation), documented assumptions, and historical cost estimates and actual experiences on other comparable programs. Be checked for accuracy, double counting, and omissions. Be updated to reflect any changes.	Partially met
Credible	Discuss any limitations of the analysis because of uncertainty, or biases surrounding data or assumptions. Risk and uncertainty analysis should be performed to determine the level of risk associated with the estimate Have results that are cross-checked against an independent estimate.	Partially met

● Met
◕ Substantially met
◑ Partially met
◔ Minimally met
○ Not met

Source: GAO analysis of NASA data.

Note: Met: NASA provided complete evidence that satisfies the entire criterion; Substantially Met: NASA provided evidence that satisfies a large portion of the criterion; Partially Met: NASA provided evidence that satisfies about half of the criterion; Minimally Met: NASA provided evidence that satisfies a small portion of the criterion; and Not Met: NASA provided no evidence that satisfies any of the criterion. See appendix III for our evaluation of JWST's cost estimate process.

Figure 3. Summary of Results of GAO Assessment of JWST Cost Estimate Based on Best Practices Criteria.

For example, as part of its cost estimation process, the project conducted a joint cost and schedule risk analysis, or joint cost and schedule confidence level (JCL), which assigned a 66 percent confidence level to the estimate.[6] In addition, we found that the cost estimate included all life cycle costs for the project. Although NASA's methods for developing the JWST cost estimate reflect some features of best practices, our review of the estimate showed that based on best practice criteria, it did not fully meet the four characteristics of a reliable estimate. See figure 3.

Specifically, the project's estimate was found to substantially meet the best practice criteria for being comprehensive, and the remaining three characteristics of being well documented, accurate, and credible were found to be only partially met. For example, the accuracy of the cost estimate, and therefore the confidence level assigned to the estimate, was lessened by the schedule used in the JCL analysis because it prevented us from, among other things, identifying the activities that were on the critical path—defined as time associated with activities that drive the overall schedule. The credibility of the estimate was lessened because project officials did not perform a sensitivity analysis that would have identified key drivers of costs, such as workforce size. Although NASA is not required to adhere to these best practices, our prior work has shown that not following best practices for cost estimating can make the cost estimate less reliable, putting projects at risk of experiencing cost overruns, missed deadlines, and performance shortfalls.[7] The best practices stem from practices federal cost estimating organizations and industry use to develop and maintain reliable cost estimates, including the Department of Defense and NASA. According to program officials, it would have been difficult, if not impossible, for the project to have met all of the best practice criteria given the complexity of the project and that some elements of the project are quite mature in their development. Instead, the program manager stated that the project followed a tailored process to develop the cost estimate that was appropriate for the project. Furthermore, officials report the project is currently meeting a majority of its milestones and executing as planned to the revised estimates for the JWST.

Comprehensive: The JWST cost estimate substantially met the criteria for being comprehensive because it included all life-cycle costs, documented cost-influencing ground rules and assumptions, and used a work breakdown structure, or the structure used to define in detail the work necessary to accomplish its objectives, that is detailed and traceable to the cost of each work element and the contract statement of work. cost estimate is not considered fully comprehensive, however, because we were unable to align the schedule to the cost estimate and could not align all of the subcontractor work to the work breakdown structure of the cost estimate. Although NASA officials believe that the project has visibility into, and can map, subcontractor activities, we were unable to confirm that cost elements were neither omitted nor double counted. In addition, we were unable to map the project's earned value management data to the work breakdown structure because information between the two was not compatible.[9] Finally, although the project outlined and documented the ground rules and assumptions, we were unable to determine whether risks associated with any assumptions were identified and traced to specific elements.

Well documented: The JWST cost estimate only partially met the criteria for being well documented because it did not include a step-by-step description of how the estimate was developed, the raw data used to develop the estimate, or the calculations and estimating methodology for specific cost elements of the work breakdown structure. Without good documentation, a cost analyst unfamiliar with the program will not be able to replicate the estimate, because he or she will not understand the logic behind it. Good documentation, for example, assists management and oversight in assessing the credibility of the estimate, helps to keep a history of reasons for cost changes and to record lessons learned, defines the scope of the analysis, and answers questions about the approach or data used to create the estimate. Project documentation, however, does provide evidence that NASA management reviewed and accepted the cost estimate because managers were briefed on the technical aspects of the estimate and were provided an overview of the joint cost and schedule risk analysis that was conducted.

Accurate: The JWST cost estimate only partially met the criteria for being accurate because the projected costs of schedule reserve did not reflect actual data, the summary schedule used to derive the JCL prevented us from sufficiently understanding how risks were incorporated, and the project did not provide evidence that it regularly updates the estimate or plans to conduct another JCL. For example, using historical actual cost data from Northrop Grumman, we estimated that 13 months of schedule reserve is likely to be $204 million instead of NASA's estimate of $121 million—a potential underestimation of 69 percent related to the schedule reserve. Project officials, however, believe they have adequate reserves available to offset any underestimation. In addition, the summary schedule the project used as an input to the JCL, although deemed acceptable by NASA, contained many long-duration activities, some with 1,000 days or more. Because of these long durations in the summary schedule used for the JCL, the lack of detail prevented us from identifying the activities that were on the critical path, as well as which risks were applied to remaining activities.[10] As a result, there is no way to ensure that risks were appropriately assigned to activities in the schedule to account for the impact of the risks during the JCL analysis.

Finally, it was unclear whether the cost estimate was regularly updated to reflect material changes in actual costs and in the project itself, such as when schedules or other assumptions change, due to a lack of detailed documentation for the cost estimate. Project officials stated that in keeping with NASA policy they do not plan, nor are they required, to conduct another JCL analysis. GAO's cost estimating best practices call for estimates to be continually updated through the life of the project, ideally every month as actual costs are reported in earned value management reports, and that a risk analysis and risk simulation exercise—like the JCL analysis—be conducted periodically through the life of the program, as risks can materialize or change throughout the life of a project. Unless properly updated on a regular basis, the cost estimate cannot provide decision makers with accurate information to assess the current status of the project. NASA officials state that the life-cycle cost estimate is updated annually for the budgeting process, and that historical records such as earned value data were used to develop the estimate. They also stated that this information is updated in several different documents being provided to management; however, we were unable to determine how this information was used in updating the cost estimate on a regular basis.

Credible: The JWST cost estimate only partially met the criteria for being credible because project officials did not adequately test and verify the reasonableness of the cost estimate and the schedule used in conducting the JCL did not have a valid critical path and contained durations that were too long to properly account for risks. For example, project officials said they did not perform a sensitivity analysis for the cost estimate. A sensitivity analysis identifies key elements that drive cost and permits analysis of different outcomes and is often used to develop cost ranges and risk reserves. NASA officials stated that the largest cost driver for the JWST project is the size of the workforce, which could have been subjected to a sensitivity analysis; yet, the cost model did not include a sensitivity analysis that would show how staff increasing or decreasing over time affects cost. In addition, NASA officials believe that all risks were sufficiently accounted for when conducting the JCL, however, the software used to conduct the JCL analysis does not recognize certain risks that officials had placed on activities in the project schedule and, therefore, some risks were discarded during the simulation.

The schedule used to conduct the JCL was also summarized at such a high level that the durations were too long to effectively model the risks. For example, one of the activities that drove the launch date was over 4 years in duration and should have been broken down further prior to conducting the simulation. Moreover, the critical path in the JCL schedule consisted of six level of effort activities all with the same duration of 2,238 days in length.[11] Level of effort activities should never be on the critical path because support activities should never drive any milestone finish date. As a result of the schedule used in the JCL not fully meeting best practices, we question the results of the analysis. Furthermore, the risk of having to carry the JWST workforce to support the project if delayed was not included since a sensitivity analysis was not performed. Project officials report that, instead, risk associated with the workforce was factored in when establishing cost reserves.

In addition, project officials did not commission an independent cost estimate, which is considered one of the best and most reliable estimate validation methods because it shows whether other estimating procedures produce similar results, and it provides an independent view of expected program costs that tests the program office's estimate for reasonableness. 12 An estimate that has not been reconciled with an independent cost estimate has an increased risk of being underfunded because the independent cost estimate provides an objective and unbiased assessment of whether the project estimate can be achieved. Notably, however, project officials provided evidence that an independent cost assessment was done for the project at the request of the JWST Standing Review Board, the independent review team for the project, and the assessment was within 2 percent of the project's estimated cost for the rebaseline. Project officials contend that the approach they used in developing the life-cycle cost estimate for the project is more accurate than the types of approaches often used to develop and independent estimate.

Complete Schedule Not Finalized; Full Assessment of Schedule Needed

We did not conduct a full schedule assessment to determine the reliability of the revised schedule based on best practices due to on-going contract negotiations.[13] The project has an integrated master schedule developed as part of the replan; however, it is not finalized because major contract modifications have yet to be negotiated and definitized.[14] Specifically, the modification to the Northrop Grumman contract, which accounts for approximately 40 percent of the total project cost and spans much of the work on the spacecraft and OTE, remains undefinitized more than a year after the project was rebaselined. Once the project completes negotiations for the contract modification and all schedule dates are set, the project can then have a measurable integrated master schedule. Project officials stated that the negotiation process and updating of associated schedules are planned to be complete in January 2013 for the Northrop Grumman contract modification—a year after submission of the latest update to its proposal for the replan. The project also reported that multiple audits of the proposals submitted by Northrop Grumman and its subcontractor by the Defense Contract Audit Agency have delayed definitization. Negotiations for the modification to NASA's contract with the Space Telescope Science Institute to incorporate the October 2018 launch readiness date are not scheduled to be complete until spring 2013.

Once all the contracts have been definitized and the project's integrated master schedule is baselined, we plan to conduct a comprehensive best practices assessment of the reliability of the project's schedule estimates.

TECHNICALLY CHALLENGING JWST PROJECT LACKS THE SCHEDULE RESERVE FLEXIBILITY AND COMMENSURATE OVERSIGHT OF INTEGRATION AND TEST EFFORTS

Project officials report that the JWST schedule has 14 months of reserve, which meets Goddard guidance for schedule reserve; however, only 7 of the 14 months are likely to be available for the last three of JWST's five complex integration and test efforts. GAO's prior work shows that it is during integration and test where problems are commonly found and schedules tend to slip. Given that JWST has a challenging integration and test schedule, this could particularly be the case. The project has made some significant progress in the past year, notably successfully completing development of the 18 primary mirror segments—considered JWST's top technical risk. Nevertheless, ongoing challenges are indicative of the kinds of issues that can require a significant amount of effort to address. For example, instrument challenges have delayed the first integration and test effort. In addition, key long-term risks on subsystems with a significant amount of work remaining will not be retired until 2016. Currently, NASA's plan for project oversight calls for one independent system integration review about 13 months before launch. While this is consistent with what NASA requires for its projects, this approach may not be sufficient for a project as complex as JWST. As a result, the current plan may be inadequate to ensure key technical and management issues are identified early enough to be addressed within the current integration and test phase schedule.

Test and Integration Schedule Lacks Schedule Reserve Late in the Process

JWST has a complex and lengthy integration and test phase, which includes five major integration and test efforts—ISIM, OTE, OTIS, spacecraft, and observatory. See figure 4 for the project reported dates for the major integration and test efforts and the schedule reserve allocated for each effort.

Overall, project officials report that the critical path schedule has 14 months of reserve with 7 months after the ISIM and OTE integration and test efforts. If these efforts are delayed beyond those 7 months, they will impinge on the schedule for the remaining three integration and test efforts. Project officials stated that the baseline plan is for the OTIS integration and test effort to not begin earlier than May 2016. These officials reported it is likely that all of the 7 months of schedule reserve held by the OTE subsystem will be utilized during its integration and test prior to delivery to OTIS and that the OTE effort is on the critical path for the project. Therefore, the remaining integration and test efforts—OTIS, Spacecraft, and Observatory—will likely have at most 7 months divided among them to use if issues are found during integration and test.

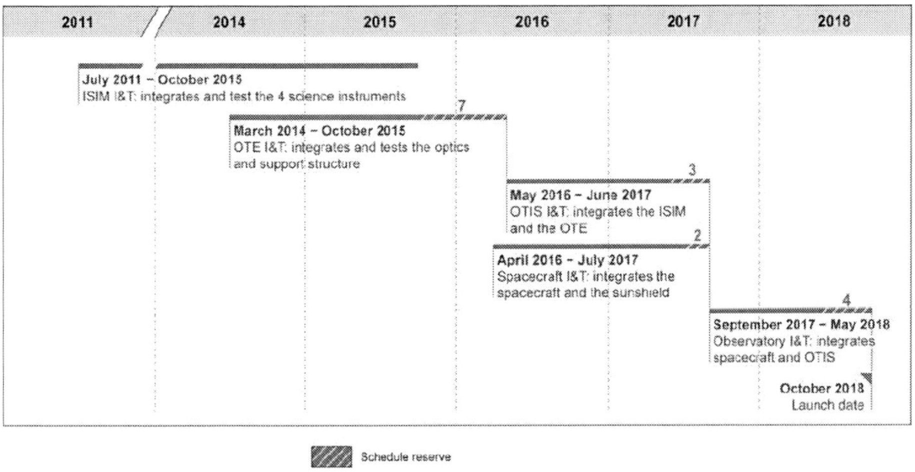

Source: GAO analysis of NASA data.

Figure 4. Project Integration and Test Time Line.

In addition to not likely being able to conserve any of the unused first 7 months of schedule reserve, the project has limited time allocated to the final three integration and test efforts, with between 2 to 4 months for each. This time could be used easily by the project if an issue were to arise during integration and test. An example of this is seen in the OTIS integration and test schedule, which currently has 3 months of schedule reserve. The final event in the OTIS integration and test effort is a lengthy cryo-vacuum test—the first time that the optics integrated with the instruments will be tested at operational temperatures near absolute zero (less than -400 degrees Fahrenheit)—that takes approximately 3 months, due to the requirements of the test. If an issue were to arise during this test that requires shutting the test down and working on the hardware, the chamber would have to be slowly warmed to a temperature safe for removal of the hardware from the chamber, work would be performed, and the 3-month test process would need to begin again. This could easily exhaust the available schedule reserve. Prior GAO work shows that it is during integration and test when problems are commonly found, and schedules tend to slip. A project official confirmed that this is the case because during integration and test the process is more sequential and there is less flexibility to move work around if problems are found. A NASA Inspector General report on the Mars Science Laboratory, another complex and high-cost mission, found that historically the probability that schedule-impacting problems will arise is commensurate with the complexity of the project.[15] JWST is one of NASA's most technologically complex projects to date.

Despite Overcoming Several Technical Challenges, JWST Has a Significant Amount of Work with Complex Technical Challenges Remaining

The project has made significant progress overcoming several technical challenges over the last year. In December 2011, for example, the project completed development of the 18 segments of the primary mirror—the project's primary technology risk—approximately 6 weeks ahead of schedule. In addition, project officials stated that during the last year they were also able to accelerate other optics-related work, which added one month of funded reserve to the schedule, bringing the total to 14 months. Finally, the project successfully addressed an increase in the estimated amount of heat on the instruments, which otherwise could have pushed observatory temperatures close to where the optics would not function correctly.

Although technical challenges are being overcome, the project will likely continue to experience additional challenges over the remainder of the project, given the significant portion and complexity of the work remaining. Four of six major subsystems have nearly 50 percent or more of their development work remaining based on its current budget information, although the dollar amounts associated with the work vary. See figure 5.

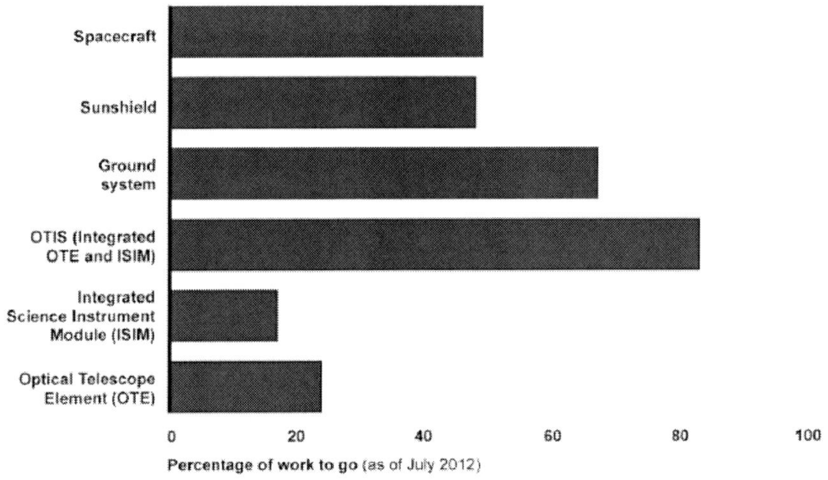

Source: GAO analysis of NASA budget data.

Figure 5. Percentage of Budgeted Development Work Remaining for Major Subsystems.

Currently, the project is experiencing several technical issues that have required a significant amount of time and effort to address. For example, the spacecraft subsystem, which experienced delays in development prior to the replan, is currently estimated to be heavier than its mass limit.[16] Spacecraft development has lagged behind other subsystems because it was viewed as a lower risk part of the project and was therefore not allocated funding when budgets were limited prior to the replan. In March 2010, the project passed its mission critical design review, which evaluated the project design and its ability to meet mission requirements and indicated that the design was ready for fabrication phase; however, the spacecraft was not included in this review due to its delayed development. Under the initial replan, which had constrained funding in fiscal years 2011 and 2012, the spacecraft critical design review was scheduled for June 2014; however, due to additional funding in the final agency-approved replan, the project was able to accelerate work and this review is now planned for December 2013.

Project officials have been concerned with the mass of JWST since its inception because of the telescope size and the limits of available launch vehicles. Accordingly, mass limits have been allocated for each subsystem, including the spacecraft. Project officials stated that they expected to encounter mass growth on the spacecraft, but that the magnitude of the mass growth on the spacecraft was unexpected. As shown in figure 6, the current spacecraft projected mass exceeds its mass allocation.

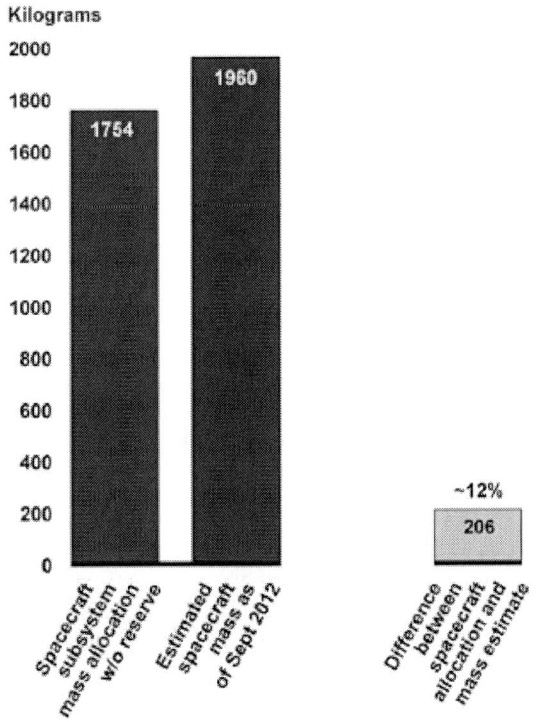

Source: GAO analysis of NASA data.

Figure 6. Current Mass Estimate for Spacecraft Subsystem.

Primary drivers of the mass growth on the spacecraft are increases in the estimated weight of the wiring harnesses, which distribute power and electric signals between different parts of the observatory, the solar array, and other structures that make up the spacecraft. The burden to find ways to reduce mass has been primarily placed with the spacecraft because it was assessed by the project to have the least technical risk and because it is the least mature subsystem and can more easily accommodate design changes. Over 100 kilograms, or 220 pounds, of mass savings options are being evaluated by the project and Northrop Grumman, which is developing the spacecraft. Potential mass solutions have been identified by Northrop Grumman and the project; however, cost and risk vary with each solution and the project is still evaluating the trade-offs of the various solutions. Project officials stated that final decisions for all tradeoffs will need to occur before spacecraft critical design review in December 2013.

The ISIM subsystem is experiencing technology and engineering challenges that resulted in the use of 18 of ISIM's 26 months of schedule reserve. The schedule for the instruments needed for ISIM continues to slip, which could result in use of more schedule reserve. Based on the replan, all four instruments were to be delivered by September 2012; however, only two instruments were delivered by that time and those still have issues that must be addressed. The remaining two instruments are currently scheduled to be delivered at least 11 months late. See table 1 below for the instrument specific issues.

Table 1. Issues with JWST Science Instruments Included in ISIM

Instrument	Instrument specific issues
MIRI	MIRI was delivered in May 2012, 1 month later than the replan date, which did not impact the ISIM integration and test schedule. However, the instrument team is still working to address an instrument sensitivity issue. The instrument team is conducting data analysis to determine a root cause for why the instrument is less sensitive thanrequired. A formal report on the root cause and recommended corrective action is expected in November 2012. NASA officials report that preliminary findings indicate that hardware changes will not be necessary.
NIRSpec	NIRSpec is currently planned for delivery in August 2013, a delay of 11 months from the replan date as a result of several technical issues. First, in July 2011, three cracks were found in the part that holds the optics components for the instrument. A failure review board was held in January 2012. ESA, which is building the instrument, will need to reassemble the instrument using the flight spare part. This part cannot be rebuilt in time to include the instrument in the first of three cryo-vacuum tests of ISIM. The project team adjusted the integration and test schedule to be able to accept delivery of NIRSpec in February 2013. However, additional issues during testing, including failure of the microshutters to close and slower than planned progress on the reintegration by the ESA contractor, have further delayed NIRSpec until August 2013. The microshutters were provided to ESA by Goddard Space Flight Center and enable the instrument to study 100 targets simultaneously. To mitigate schedule risk, the project is studying an option to build new microshutters that will replace the current ones, if necessary. Delivery of NIRSpec after August 2013 may threaten ISIM's critical path schedule.
FGS/ NIRISS	FGS/NIRISS was delivered as planned in July 2012; however, it was delivered with a known gear motor issue that could result in the early loss of NIRISS in space. In July 2012, the Canadian Space Agency (CSA), which built the instrument, began to redesign the gear motors and they are being rebuilt. Around the time of the replan, CSA had to redesign its instrument because the original design was not meeting performance requirements. Part of instrument design was simplified, which resulted in some loss of expected science performance; however, NIRISS science is not required for JWST mission success.
NIRCam	NIRCam is currently planned for delivery in August 2013, a delay of 11 months from the replan date as a result of technical and workmanship issues. The project determined the need to electrically ground a mirror on the instrument and the team is reviewing several potential options to address this issue. Workmanship issues, such aspoor soldering quality of the wires in the instrument controls electronic box, resulted in the need for corrective actions. The project has decided to run the first ISIM cryo-vacuum test without NIRCam.

Source: GAO analysis of NASA information.

In addition to the instrument delays, two other technical challenges associated with ISIM are: (1) the detectors used by three of the four instruments to capture infrared light in space are degrading and may need to be replaced, resulting in the addition of another round of cryo-vacuum testing—in which a test chamber is used to simulate the near absolute zero temperatures in space, and (2) issues with the development of the cryo-cooler system that removes heat and cools MIRI. In December 2010 the project became aware that the detectors in three of the instruments were degrading.[17] As a result, approximately $42 million and 15 months of schedule reserve to replace the detectors were included in the replan. These additions covered the cost of manufacturing the detectors; fabrication, assembly, and test of new focal plane assemblies; changing the detectors on three instruments, and the addition of a third ISIM cryo-vacuum test. The manufacturing process for new detectors takes approximately 30 months, which means that they cannot be delivered until after the second round of ISIM cryo-vacuum testing in 2014.

As a result, $2 million of the $42 million in the replan was used to add a third round of cryo-vacuum testing for ISIM. The third test will validate the performance requirement of the ISIM and is the only time the instruments are tested with the flight detectors. Changing the detectors requires disassembling the instruments from ISIM, a process that will risk damage to the structure and instruments. Project officials stated that they will continue to monitor the degradation rate of the current detectors because if the degradation rate is low, they may not replace the detectors.

Development issues with a part of the cryo-cooler needed for MIRI have delayed its delivery to ISIM. In 2010, project officials realized that an essential valve in the cryo-cooler was leaking at rates that exceeded requirements. Following the results of a failure review board, the contractor manufactured a newly designed valve, but it also did not meet leak rate requirements. Project officials stated that a new valve design will not be manufactured in time for use in the first ISIM cryo-vacuum test. The project is concurrently developing three alternatives and authorized manufacturing for one of the alternatives in October 2012. Project officials stated that the MIRI cryo-cooler is particularly complex because it spans approximately 10 meters—or approximately 33 feet—through the entire JWST observatory. These issues combined required the use of 18 months of schedule reserve, which reduced ISIM's schedule reserve from the 26 months established in the replan to 8 months before it is needed for integration with the OTIS subsystem.

These types of issues are not uncommon among NASA programs as technical issues tend to arise when disparate parts are integrated and tested together for the first time. Given the complexity and cutting edge technology developed and used on JWST, it is expected that these kinds of issues will continue to materialize as the program moves through its complex integration and test program. Figure 7 shows the delay of instrument deliveries as well as changes to the ISIM integration and test and final delivery dates over the last year.

Until the project is able to overcome the major issues with the instruments and other parts of the ISIM, it is likely that the schedule would continue to slip and may begin to affect the overall project schedule. ISIM still has 8 months of schedule reserve before the slipping of its schedule would affect the schedule for the remainder of the project. The instrument, detector, and cryo-cooler issues have all contributed to the delay in the ISIM integration and test schedule and the reduction of objectives that can be achieved in the first two rounds of cryo-vacuum testing. The first round of testing will not include two instruments, a final design of the cryocooler hardware, or new detectors. As a result project officials will only be able to gather risk reduction information on the FGS/NIRISS, MIRI, test procedures, and test support equipment from the first cryo-vacuum test.

The project also has several known long term risks and challenges remaining. For example, risks related to OTIS, the sunshield, and the ground system subsystems are not scheduled to be addressed until late in project development. As of October 2012, seven of the top 10 project risks were related to the long-term risks associated with the OTIS and sunshield, most of which will not be resolved until 2016 or later. For example, several risks relating to OTE are not scheduled to be closed until the OTIS testing in the chamber at

Johnson Space Center in February 2017. Project officials are adding risk mitigation through early and additional testing, where possible, to these subsystems. Prior to the replan, the ground system software was at high risk for not being completed before launch and many tasks were planned for completion after launch. Space Telescope Science Institute officials stated that the replan allows them to plan for completion of their work before launch on a more realistic time schedule, which decreases schedule and operational risk. A continuing challenge on the ground system is that some development and testing is dependent on the final design of subsystems such as the instruments, which continue to slip delivery dates.

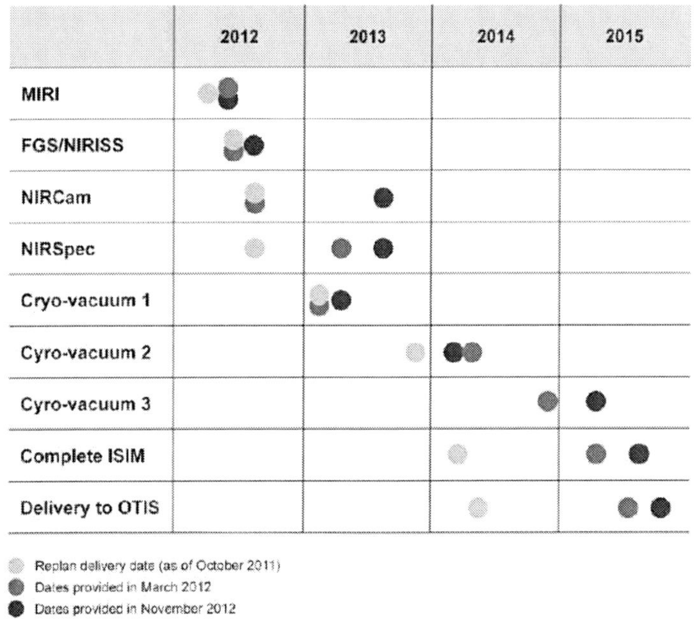

Source: GAO analysis of NASA data.

Figure 7. Changes in ISIM Schedule Since the Project Replan.

Independent and Management Oversight during Test and Integration May Not Be Commensurate with JWST Test and Integration Complexity

The project plans to hold independent and management reviews required for all projects during the integration and test phase, but this phase for JWST is particularly complex. JWST has five major integration and test efforts that span 7 years and only one independent mission-level technical review—the system integration review. The system integration review evaluates the readiness of the project and associated supporting infrastructure to begin system assembly, integration, and test, and evaluates whether the remaining project development can be completed within available resources. For JWST, this review is scheduled in September 2017, only 13 months prior to launch. Projects we reviewed that had recently launched, however, held their system integration review on average approximately 22 months prior to launch. The project has an internal review with participation from standing review board members planned before the beginning of OTIS integration and test activities begin, and it will be subject to independent lower level reviews conducted by the Goddard Systems Review Office of the integration and test process. In addition, key decision point D (KDP-D)—when the senior agency decision authority would approve the project to proceed into the system integration and test phase—is scheduled for December 2017, 3 months after the commencement of the final major integration and test activity. According to NASA policy, this review should be held prior to the start of the system integration and test phase of the project.[18] Project budget information shows that over 90 percent of expected integration and test funding will be spent on four major integration and test activities prior to the scheduled mission-level system integration review and KDP-D approval by NASA senior management. As a result, the current plan may be inadequate to ensure that key technical and management issues are identified early enough to be addressed within the current integration and test phase schedule.

JWST Project Has Taken Steps to Enhance Communications with and Oversight of Its Contractors

The JWST project has taken steps to improve communications and oversight of its contractors as part of the replanning activities. For example, based on recommendations from the ICRP, the project has instituted meetings at various levels throughout NASA and its contractors and subcontractors. In addition, the project has added personnel at contractor facilities, which has allowed for more direct interaction and quicker resolution of issues.

The project also assumed responsibility of the mission-level systems engineering function from Northrop Grumman, a move that shifts the authority to make trades or decisions to NASA. An independent NASA review of the project conducted in May 2012 found, however, that agencywide reductions in travel budgets have put the effectiveness of the JWST project's oversight plans in jeopardy. While the project received partial relief from travel budget reductions in fiscal year 2012, project officials are concerned that the current level of oversight will not be sustained if similar cuts in travel funding occur in future years as anticipated. The project is also taking steps to enhance its oversight of project risks by implementing a new risk management system. The new project manager found that the previous system lacked rigor and was relatively ineffective for managing project risks, especially for a project as complex as JWST. The new system should allow for better tracking of risks than did the previous system. While these enhancements to the oversight of the project are steps in the right direction, it will take time to assess their effectiveness.

NASA Responded to Recent Independent Review Panel Report with Increased Communication and Oversight

Based on recommendations in the ICRP report, NASA has taken action to enhance oversight and communications. See table 2 for the ICRP recommendations and actions taken by NASA in response.

Table 2. ICRP Recommendations Concerning Oversight and Communications and the Actions Taken by NASA

ICRP recommendation	Action taken
Move the JWST management and accountability from the Astrophysics Division to a new organizational entity at NASA headquarters that has responsibility only for the management and execution of JWST.	Completed: The management of JWST at Headquarters has been reorganized. JWST is now a stand-alone program reporting to the NASA Associate Administrator for programmatic oversight and to the Associate Administrator for the Science Mission Directorate for technical and analysis support. A similar approach was taken in the past with both the Hubble Space Telescope and the Mars Exploration Program.
Assign management and execution responsibility for the JWST Project to the Goddard Center Director, with accountability to the Science Mission Directorate Associate Administrator at NASA headquarters.	Completed: NASA has reorganized the JWST project at Goddard to report directly to the Center Director, who is responsible for the management and execution of the JWST Project and accountable to the NASA Administrator.
Assign at least one senior Goddard project person to be resident at Northrop Grumman throughout the Project.	In progress: In August 2012, the Deputy Observatory Manager began spending at least half time at the Northrop Grumman facility, with a permanent move scheduled in early 2013.
Improve communications between the JWST Project and both Goddard management and NASA headquarters Science Mission Directorate.	Completed: The JWST project office provides weekly updates of risks and new issues to the Goddard Center Director. In addition, the Project Manager at Goddard has daily interaction with the Program Manager at NASA HQ.
Conduct monthly or bi-monthly JWST Project Executive meetings, attended by the NASA Associate Administrator and the President of Northrop Grumman Aerospace Systems.	Completed: Meetings were first held in December 2010 between the NASA Associate Administrator, the new JWST Program Director, and the Goddard Center Director and senior management of the prime contractor. Since that time, senior level management reviews have occurred on a quarterly basis and are open to subcontractors as well. Topics at these meeting have included issues such as the spacecraft mass problems, instrument delivery delays, and the process to find alternative solutions for the MIRI cryo-cooler valve problem.
Strengthen the role and the independent voice of the science team in the Project.	Completed: The project added a Deputy Senior Project Scientist/Technical position to theproject science team. This individual: • is responsible for day-to-day interactions with senior project management on all aspects of the mission; scientific, technical, budgetary, and schedule. • meets regularly with other members of the project science team to ensure better communication with other functions in the project. • assists the Senior Project Scientist to facilitate better integration of the science activities with the hardware development activities and enable closer coordination and understanding of technical drivers to science performance so fully informed decisions can be made.

Source: GAO analysis of NASA data.

NASA has taken steps to increase communication between the project and its contractors and subcontractors in an effort to enhance oversight. According to project officials, the increased communication has allowed them to better identify and manage project risks by having more visibility into contractors' activities. The project reports that a great deal of communication existed across the project prior to the ICPR and replan; however, improvements have been made.

For example, monthly meetings between project officials at Goddard and all of the contractors have continued on a regular basis and include half-day sessions devoted to business discussions. The project reports that these meetings have benefits over other forms of communication. For example, it was through dialogue with several technical leads at Northrop Grumman during detailed reviews of analytical models that the project identified that the mass issue on the spacecraft was likely to occur.

In addition, the project has increased its presence at contractor facilities as necessary to provide assistance with issues. For example, the project has had two engineers working on a recurring basis at Lockheed Martin to assist in solving problems with the NIRCam instrument. The ISIM manager said that these engineers have insight into Lockheed Martin's work and are having a positive effect as they offer technical help and are involved in devising the solutions to issues.

He added that that these engineers have authority to make decisions on routine issues to allow the work flow to continue, but decisions that are more complex or require a commitment of funds are communicated to project management for disposition. The project reports that the Jet Propulsion Laboratory, responsible for NASA contribution to the MIRI instrument and its associated cryo-cooler, has an in-house representative in the responsible Northrop Grumman division to monitor the work being performed on the cryo-cooler.

The JWST project also assumed full responsibility for the mission system engineering functions from Northrop Grumman in March 2011. NASA and Northrop Grumman officials both said that NASA is better suited to perform these tasks. Project officials stated the systems engineering requires the ability to make trades and decisions across the entire observatory, and because Northrop Grumman is only responsible for portions of the observatory, it did not have the authority to make trades or decisions for areas outside of its control.

Although responsibility for the overall mission systems engineering function was removed from Northrop Grumman, it retains system engineering responsibility for work still under its contract, such as development of the spacecraft and sunshield. The ICRP noted that a highly capable, experienced systems engineering group is fundamental to project success and appropriate to ensure accountability especially for a project of JWST's complexity and visibility.While these enhancements to the oversight of the project are steps in the right direction, it will take time to assess their effectiveness. In addition, sustainment of these efforts on the part of the project will be important.

Project and contractor officials we spoke with believe that the increased communication has had a positive effect on the relationships between them. We will continue to monitor the interaction between the project and its contractors and its frequency in future reviews to identify whether the changes have had the desired results.

Travel Budget Reductions May Hamper Planned Oversight Activities

The JWST project reported that its travel budget was reduced by approximately $200,000 from the $1.2 million planned in fiscal year 2012 as a result of NASA's implementation of an Executive Order to promote more efficient spending.[19]

According to project officials, the changes in oversight necessitated by a reduction in travel funds represent a major shift away from the management paradigm adopted during the replan. Proposed reductions in future fiscal years could significantly reduce the project's travel budget. The project reports that the travel requirements for fiscal years 2013 through 2015 are $1.6 million, $1.7 million, and $1.8 million, respectively. Officials reported that while travel is a small percentage of the project's annual budget, the majority of expected travel—about 87 percent—is for oversight functions put in place as a result of the ICRP recommendations, such as having a permanent on-site presence at Northrop Grumman. These oversight functions include attending and participating in contractor monthly programmatic and technical reviews, technical interface meetings, recurring on-site presence at contractor facilities for quality assurance reviews and inspection of hardware. JWST project officials are concerned that decreased oversight could translate into the project increasing its use of cost and schedule reserves as they will not be conducting planned oversight to better ensure success.

A recent NASA Office of Evaluation review concluded that by not having an adequate travel budget, the project is at risk of cost/schedule growth and/or technical risk due to the late identification of issues or timely resolution strategies.

The project has made adjustments to absorb the reduction in fiscal year 2012 and plans to identify instances of increased cost or schedule risk due to late identification of issues. However, the project does not have a strategy to address anticipated future reductions. Ensuring adequate oversight is particularly important as the project begins its complex and lengthy test and integration phase, where issues will likely surface.

NASA Headquarters Independent Review of the JWST Program Identified Concerns and Recommends Indicators to Measure Progress

As part of NASA's approach to increase oversight of the project at headquarters, NASA's Office of Evaluation recently conducted an independent review of the JWST project to assess the progress since the September 2011 rebaseline was approved. According to the Director of the Office of Evaluation, the goal of the review was not to reproduce the replan assessment, but rather to assess progress based on cost, schedule, and technical performance of the project and the status of oversight functions within NASA headquarters, the JWST Program Office, and Goddard Space Flight Center. The intended outcome of the review was 1) to obtain a snapshot of performance to determine if the program was progressing in accordance with its plan, and 2) to identify leading indicators for upper management to use when tracking future performance. The review team identified several areas of concern within the program, many of which we have highlighted, and recommended a list of leading indicators that project management should consider tracking.

The Director of the Office of Evaluation said that the project is generally performing the activities and maintaining the schedule set forth in the replan; however, the team identified key areas that should be monitored as the project moves forward. The review team also recommended a set of leading indicators for project management to consider tracking to measure and monitor progress. The Director added that these indicators are for the project to use and would not be specific criteria for use by independent review boards such as the Standing Review Board. These indicators are a positive step to ensure that NASA management has the information necessary to monitor the progress of the JWST project. See table 3 below for the concerns raised by the review team.

Table 3. NASA Office of Evaluation Review Team Concerns

Concern identified	NASA Office of Evaluation Review Team finding
Resolving the on-going mass issue.[a]	The technical and programmatic risks involved in possible solutions need to be explicitly understood.
Understanding root causes for the NIRSpec Optical Bench cracking issue.	There needs to be a better understanding of the root causes and investigate what would constitute a flyable structure.
Monitoring of schedule erosion by the project and recognition of the dynamics of four competing critical paths.[a]	There are several competing critical paths for the JWST program to include development and integration and test of the OTE, ISIM, sunshield, and the spacecraft.
Monitoring the stability of the schedule going forward.	The master schedule took 8 months to baseline and stability of the schedule going forward is unknown. The project will need an accurate reporting system at the integrated master schedule level.
Ensuring that Northrop Grumman will have an adequate workforce and that skills are available in fiscal year 2013.[a]	The project needs to work with Northrop Grumman to ensure that it has an adequate workforce and skill sets available to accomplish the planned ramp up for fiscal year 2013.
Monitoring the increased risk of cost/schedule growth and technical risk.	Without an adequate travel budget to execute the original plan, there is an increased risk ofcost or schedule growth and technical risks due to late identification of issues or timely resolution strategies.

Source: GAO analysis of NASA data.

[a]These issues were also identified as leading indicators the review team recommended that project management should consider tracking.

Changes Made to Project Risk Management System

The new JWST project manager re-emphasized the importance of the project's risk management system and, in August 2012, a new risk management database was implemented to support the system. The project manager told us that he evaluated the risk management system being utilized by the project when he assumed his position and found it to be ineffective and not robust, especially for a project as complex as JWST. While the basic risk management methodology remains unchanged, the project manager wanted a more regimented system. For example, the project utilizes a hierarchy of risk boards that periodically reviews and provides disposition of all new and existing risks. These risk boards reviewed and assessed new risks and lower level risk board actions and met on an ad hoc basis. The project manager instituted a more regimented system that re-emphasized and revised the weekly project risk board meetings. Lower level risk boards meet a minimum of once a month depending on activity.

The project manager also determined that a new risk management database needed to be put in place that would bring more rigor to the risk management process. The project manager told us that he directed an overhaul of the risk management database to provide more complete information to management on the purpose and history for each risk. The goal was to improve consistency in how the project determined the potential for a risk to occur and its impact, and provide greater detail on mitigation and better tracking of the status for each risk.

For example, the new system puts more emphasis on understanding and capturing the key events in the mitigation plan that are intended to result in a change in likelihood or consequence of a risk. The new system has a provision where the mitigation plan will be entered and updated over time, and the capability to store data such as mitigation steps throughout the life of the risk. In addition, the new system now archives data automatically to provide a traceable history of the risk. The prior data system did not have as robust of an archiving function. Furthermore, the project manager wanted to improve the linkage between the risk database entries and financial records to ensure consistency of the data in the risk database with regard to cost and schedule for risk mitigations with project office financial records. As the changes to the risk management system and database, as well as other changes we identified that were put in place to enhance oversight were just recently implemented, we will continue to monitor their continued use and assess the impact they may be having on the project.

CONCLUSION

The JWST project is among the most challenging and high-risk projects NASA has pursued in recent years. It is also one of the most expensive, with a recent major replan resulting in a total cost of $8.8 billion. The reasons for cost and schedule growth were largely recognized by an independent review team to be rooted in ineffective funding, management, communication, and oversight. NASA has invested considerable time and resources replanning the project and instituting management and oversight improvements in order to ensure that it (1) can be executed within its new estimates and (2) has addressed the majority of issues raised in the recent independent review. It appears that communications with contractors and within NASA have improved, that a more robust risk mitigation system is in place, that more is known about what it will take to complete the project and how much it will cost, and that the project is currently meeting the majority of its milestones. Nevertheless, over the course of the next several years, the project will be executing a large amount of work with several extremely complex and challenging integration and test efforts. Because three major test and integration efforts must be completed in the last 2 years of the JWST schedule, it is essential that issues are identified and addressed early enough to be handled within the project's current schedule. While the JWST oversight plan is consistent with NASA's requirements for all project's required reviews, a single independent review scheduled just over a year before launch may not be sufficient to identify and resolve problems early for a project of this magnitude. A key element of overseeing project progress is monitoring how the project is executing to its cost baseline. To that end, while NASA took some steps that were in line with best practices to develop its revised baseline, some of the deficiencies we found in its process could impact the reliability of the cost estimate and the joint cost and schedule confidence level that was provided to headquarters decision-makers. Without higher-fidelity, regularly updated information related to costs, as well as an oversight regime during later phases of test and integration that is commensurate with the complexity of that effort, NASA risks late identification of technical and cost issues that could delay the launch of JWST and increase project costs beyond established baselines. Also important to oversight for the remainder of the project is the ability of officials to sustain improvements to communication with and oversight of contractors. Anticipated travel restrictions, however, could decrease the project team's ability to sustain these actions. Without a plan to address such reductions in future years, the project could once again become susceptible to communication and oversight problems identified in earlier reviews, which could also have a detrimental impact on continued project performance.

RECOMMENDATIONS FOR EXECUTIVE ACTION

To ensure that the JWST life-cycle cost estimate conforms to best practices, GAO recommends that the NASA Administrator direct JWST officials to take the following three actions to provide high-fidelity cost information for monitoring project progress:

- improve cost estimate documentation and continually update it to reflect earned value management actual costs and record any reasons for variances,
- conduct a sensitivity analysis on the number of staff working on the program to determine how staff variations affect the cost estimate, and
- perform an updated integrated cost/schedule risk analysis, or joint cost and schedule confidence level analysis, using a schedule that meets best practices and includes enough detail so that risks can be appropriately mapped to activities and costs; historical, analogous data should be used to support the risk analysis.

To ensure that technical risks and challenges are being effectively managed and that sufficient oversight is in place and can be sustained, GAO recommends that the NASA Administrator direct JWST officials to take the following three actions:

- conduct a separate independent review prior to the beginning of the OTIS and spacecraft integration and test efforts to allow the project's independent standing review board the opportunity to evaluate the readiness of the project to move forward, given the lack of schedule flexibility once these efforts are under way,
- schedule the management review and approval to proceed to integration and test (key decision point D or KDP-D) prior to the start of observatory integration and test effort, and
- devise an effective, long-term plan for project office oversight of its contractors that takes into consideration the anticipated travel budget reductions.

AGENCY COMMENTS AND OUR EVALUATION

NASA provided written comments on a draft of this report. NASA also provided technical comments, which were incorporated as appropriate.

In responding to a draft of this report, NASA concurred with three recommendations and partially concurred with three other recommendations and commented on actions in process or planned in response. In some cases, these actions meet the intent and are responsive to issues we raise; however, some of the responses do not fully address the issues we raised in the report.

NASA partially concurred with our recommendation to improve the cost estimate documentation of the JWST project, and to continually update it to reflect earned value management actual costs and record any reasons for variances between planned and actual costs. In response to this recommendation, NASA officials stated that the project currently receives earned value data from some of its contractors and performs monthly analysis of that data to understand the contractors' estimates at completion, and then compares these numbers to similar figures independently assessed by the JWST project. NASA also highlighted its efforts to improve the agency's documentation of the earned value variances and to extend the earned value management analysis to areas where it is not yet implemented, such as ground systems development at the Space Telescope Science Institute. In addition, NASA responded that its annual budget process generates a requirements-driven budget plan consistent with the rebaseline. NASA stated that this information is updated in several different documents that are provided to management and it does not plan to revise its JCL documentation developed during the replan. Despite these steps, we could not independently confirm that they were leading to an updated cost estimate, which is the basis of our recommendation. If the estimate is not updated, it will be difficult to analyze changes in project costs and collecting cost and technical data to support future estimates will be hindered. Furthermore, if not properly updated on a regular basis, the cost estimate cannot provide decision makers with accurate information for assessing alternative decisions. Without a documented comparison between the current estimate (updated with actual costs) and the old estimate, the cost estimator cannot determine the level of variance between the two estimates and cannot see how the project is changing over time. Therefore, we continue to believe NASA will be well served by following best practices and updating its cost estimate with current information and documenting reasons for any variances. We encourage the project to improve the cost estimate documentation and record any reasons for variances between planned and actual costs and we intend to review the documentation as a part of our ongoing review of the project.

NASA officials partially concurred with our recommendation that the project conduct a sensitivity analysis on the number of staff working on the project to determine how staff variations affect the cost estimate. In its response, the agency stated that it believes it met the intent of this recommendation when staffing levels were determined in the 2011 JWST rebaseline based on programmatic experience from the accomplishment of similar activities. To accommodate the possibility of increased costs based on increased staffing hours, NASA reports that funded schedule reserve was built into the JWST rebaseline, in addition to unallocated future expenses being held at various levels of the organization. NASA believes that these reserves will be sufficient to cover increases for the duration of specific activities that result in increased staffing cost, and that an additional workforce sensitivity analysis is not warranted. NASA added that the joint cost and schedule confidence level analysis performed provided a de facto workforce sensitivity analysis and does not plan any further action. A joint cost and schedule confidence level analysis, however, is not the same as a sensitivity analysis wherein the sources of the workforce variation should be well documented and traceable. While we appreciate the steps NASA took to account for workforce variation, the JWST cost model does not show how staff levels increasing or decreasing over time affects cost. Furthermore, best practices call for a risk analysis to be conducted in conjunction with a sensitivity analysis, not to be a substitute for it. As a best practice, a sensitivity analysis should be included in all cost estimates because it examines the effects of changing assumptions and ground rules. Since uncertainty cannot be avoided, it is necessary to identify the cost elements that represent the most risk and, if possible, cost estimators should quantify the risk. Without performing a sensitivity analysis that reveals how the cost estimate is affected by a change in a single assumption, such as workforce size, the cost estimator will not fully understand which variable most affects the cost estimate. Therefore, we continue to believe that NASA should conduct a sensitivity analysis for the JWST project, given the large number of staff working on the program, to determine how staff variations positively or negatively affect the cost estimate rather than relying on schedule reserve and unallocated future expenses to offset any shortfall.

NASA concurred with our recommendation to perform an updated integrated cost and schedule risk analysis using a schedule that meets best practices and includes enough detail so that risks can be appropriately mapped to activities and costs. In response to this recommendation, NASA stated that the agency is already using tools and a method to conduct programmatic assessments of projects after the baseline was established using the JCL methodology. While these may be good tools, the key point is the need to address shortcomings of the schedule that supports the baseline itself. For example, the lack of detail in the summary schedule used for the joint cost and schedule risk analysis prevented us from sufficiently understanding how risks were incorporated; therefore, we question the results of that analysis. Since the JCL was a key input to the decision process of approving the project's new cost and schedule baseline estimates, we maintain that the JWST project should perform an updated JCL analysis using a schedule with sufficient detail to map risks to activities and costs. Doing so could help increase the reliability of the cost estimate and the confidence level of the JCL. Furthermore, risk management is a continuous process that constantly monitors a project's health. Given that JWST is many years from launch and the risks that the project faces are likely to change, a risk analysis should be conducted periodically throughout the life of the project.

NASA concurred with our recommendation to conduct a separate independent review prior to the beginning of the OTIS and spacecraft integration and test efforts. In response to this recommendation, NASA stated that it will request members of the independent JWST Standing Review Board participate in OTIS Pre-Environmental Review scheduled prior to the beginning of OTIS environmental testing. A member of the Standing Review Board will co-chair this review and report its findings to the NASA Associate Administrator, which is the practice of all Standing Review Board reviews. In addition, NASA plans to direct Northrop Grumman, the spacecraft developer, to add members of the Standing Review Board, as well as members of the Goddard Independent Review Team, to the spacecraft element integration readiness review and report their findings to the NASA Associate Administrator. We believe these actions meet the intent of our recommendation and will afford an independent evaluation of the readiness of the project to move forward with its major integration and test efforts.

NASA partially concurred with our recommendation to schedule the management review and approval to proceed to integration and test (KDP-D) prior to the start of the observatory integration and test effort. In response to this recommendation, NASA stated that it will reduce the 3- month gap between the scheduled system integration review and the KDP-D review, which it believes will provide NASA management and the NASA Associate

Administrator with the full independent assessment earlier than currently planned. While we agree that this change will move the review earlier than previously planned, based on its response, NASA still plans to hold the review after the observatory integration and test is already underway. Holding this review after the observatory integration and test effort is already underway does not meet agency policy and will lessen the impact of the review as it may be inadequate to ensure key technical and management issues are identified early enough to be addressed. KDP-D is the point in which management approval is given to transition to the test and integration phase. We reiterate our recommendation that NASA should hold this important key decision point prior to the beginning of this last major integration and test effort, as required by agency policy.

NASA concurred with our recommendation to devise an effective, longterm plan for project office oversight of its contractors that takes into consideration the anticipated travel budget reductions. In response to this recommendation, NASA stated that it will develop a plan based on fiscal year 2013 travel allocations and will take into consideration anticipated travel budget reductions. In addition, NASA stated that the plan will enable the project to maintain oversight of JWST contractors and their ability to meet performance and delivery deadlines and work closely with the international partners. We believe such a plan will be critical to ensuring adequate oversight, which is particularly important as the project enters into the complex integration and test efforts where issues will likely surface. In addition, we agree with the concerns of project officials that the current efforts to increase communication and oversight may not be sustained if reductions to future travel budgets occur as anticipated. We encourage the project to complete this plan in a timely manner and intend to review it as a part of our ongoing assessment of the project's oversight efforts.

Cristina T. Chaplain
Director
Acquisition and Sourcing Management

Appendix I: Objectives, Scope, and Methodology

Our objectives were to assess (1) the extent to which NASA's revised cost and schedule estimates are reliable based on GAO best practices, (2) the major risks and technological challenges the James Webb Space Telescope (JWST) project faces, and (3) the extent to which the National Aeronautics and Space Administration (NASA) has improved the oversight of the JWST project. In assessing the project's cost and schedule estimates, we performed various checks to determine that the provided data were reliable enough for our purposes. Where we discovered discrepancies, we clarified the data accordingly. Where applicable, we confirmed the accuracy of NASA-generated data with multiple sources within NASA.

To assess the current life cycle cost estimate of the JWST project, we used the GAO cost guide to evaluate the estimating methodologies, assumptions, and results to determine whether the cost estimate was comprehensive, accurate, well-documented, and credible.[1]

We developed standardized data collection instruments to request relevant cost and schedule documentation and questionnaires for project officials to complete. Through the data collection instruments and questionnaires, we gathered basic information pertaining to the project's cost and schedule estimation process.

We examined documents from the JWST replan efforts pertaining to the revised cost estimate and the joint cost and schedule confidence level (JCL) analysis, including detailed spreadsheets that contained cost, schedule and project risk information. Project documents we reviewed included basis of estimates for the replan from contractors and the project office, contractor engineering change proposals, the JWST program plan and mission requirements, as well as the NASA cost estimating handbook.[2] To assess how management and independent review teams were involved in the cost estimate process, we also reviewed the project's monthly status presentations to Goddard management, Standing Review Board reports concerning the project's revised baseline estimates, cost estimates conducted by NASA's Independent Program Assessment Office, and the Agency Program Management Council's decision memo for the rebaselined estimate and JCL analysis.

We supplemented our analysis by interviewing officials from Northrop Grumman, the Space Telescope Science Institute, and the JWST program and project offices.

After reviewing cost estimate documentation submitted by NASA and conducting numerous interviews with relevant sources within the project office, we calculated the assessment rating of each criteria within the four characteristics by assigning each individual assessment rating: Not Met = 1, Minimally Met = 2, Partially Met = 3, Substantially Met = 4, and Met = 5. We then took the average of the individual assessment ratings for the criteria to determine the overall rating for each of the four characteristics. The resulting average becomes the "Overall Assessment" as follows: Not Met = 1.0 to 1.4, Minimally Met = 1.5 to 2.4, Partially Met = 2.5 to 3.4, Substantially Met = 3.5 to 4.4, and Met = 4.5 to 5.0. We discussed the results of our assessments with officials within the program office at NASA headquarters and the project office at Goddard Space Flight Center.

We supplemented the assessment of the revised 2011 cost estimate with an assessment of the summary schedule used for the JCL, which was a part of the project's cost estimation process, and followed criteria laid out in the GAO schedule guide.[3] These practices address whether the schedule (1) captured all activities; (2) sequenced all activities—that is, listed in the order in which they are to be carried out; (3) assigned resources to all activities; (4) established the duration of all activities; (5) integrated schedule activities horizontally and vertically, which identifies whether products and outcomes associated with other sequenced activities are arranged in the right order, and that varying levels of activities and supporting subactivities are also aligned properly; (6) established for all activities, the critical path, which is the longest continuous sequence of activities that is necessary to examine the effects of activities slipping in the schedule; (7) identified between activities float, which is the amount of time by which a predecessor activity can slip before the delay affects the program's estimated finish date; (8) identified a level of confidence using a schedule risk analysis; and (9) was updated using logic and durations to determine dates. We also reviewed the inputs to the JCL model, the document outlining the methodology of the analysis that accompanied the electronic files, and interviewed cognizant project officials to discuss their use of the summary schedule.

Because the project's detailed integrated master schedule has not been finalized because of ongoing negotiations and contract modifications, we did not conduct a complete schedule analysis using the GAO schedule assessment guide. We plan to perform this assessment in a subsequent review of the JWST project.

To assess the major short- and long-term risks and technological challenges facing the project, we reviewed the project's risk list, monthly status reviews, and other documentation provided by projects and contractor officials. This information covered the risks, mitigation plans, and timelines for addressing risk and technological challenges.

We also interviewed project officials for each major observatory subsystems to clarify information and to obtain additional information on risks and technological challenges. Further, we interviewed officials from the Jet Propulsion Laboratory, Northrop Grumman Aerospace Systems, Lockheed Martin Advanced Technology Company, Teledyne Imaging Sensors, the University of Arizona, and the Space Telescope Science Institute concerning risks and challenges on the subsystems, instruments, or components they were developing. We reviewed GAO's prior work on NASA Large Scale Acquisitions, NASA Office of Inspector General reports, and NASA's Space Flight Program and Project Management Requirements and Systems Engineering Processes and Requirements policy documents.[4]

We compared NASA's controls as outlined in these agency policies with the project plan to assess the extent to which the JWST's plan followed the intent of the policies with regard to independent oversight and management approval processes.

To assess the extent to which NASA is performing enhanced oversight of the JWST project, we reviewed documentation from the Independent Comprehensive Review Panel and the project to determine actions taken by NASA in response to the panel's recommendations.

We interviewed project officials to understand the impact of these changes on the oversight processes for the project and communication between the project and its contractors. We also interviewed officials from the Jet Propulsion Laboratory, Northrop Grumman Aerospace Systems, Lockheed Martin Advanced Technology Company, Teledyne Imaging Sensors, the University of Arizona, and the Space Telescope Science Institute concerning project oversight of work they were performing and the effectiveness of oversight changes. In addition, we reviewed a presidential directive and Office of Management and Budget and project documentation and interviewed project officials concerning the reductions to travel budgets and their impact on project oversight activities.

We interviewed the Director of NASA's Office of Evaluation about a recent internal review of the JWST project and reviewed documentation from that review. We also reviewed documentation and interviewed project officials concerning the changes made to the project's risk management system.

Our work was performed primarily at NASA Headquarters in Washington, D.C., and Goddard Space Flight Center in Greenbelt, Maryland. We also visited Johnson Space Center in Houston, Texas, and the Jet Propulsion Laboratory in Pasadena, California.

In addition, we met with representatives from Northrop Grumman Aerospace Systems, Lockheed Martin Advanced Technology Company, Teledyne Imaging Sensors, the University of Arizona, and the Space Telescope Science Institute.

We conducted this performance audit from February 2012 to December 2012 in accordance with generally accepted government auditing standards. Those standards require that we plan and perform the audit to obtain sufficient, appropriate evidence to provide a reasonable basis for our findings and conclusions based on our audit objectives.

We believe that the evidence obtained provides a reasonable basis for our findings and conclusions based on our audit objectives.

Appendix II: Subsystems of the JWST Observatory

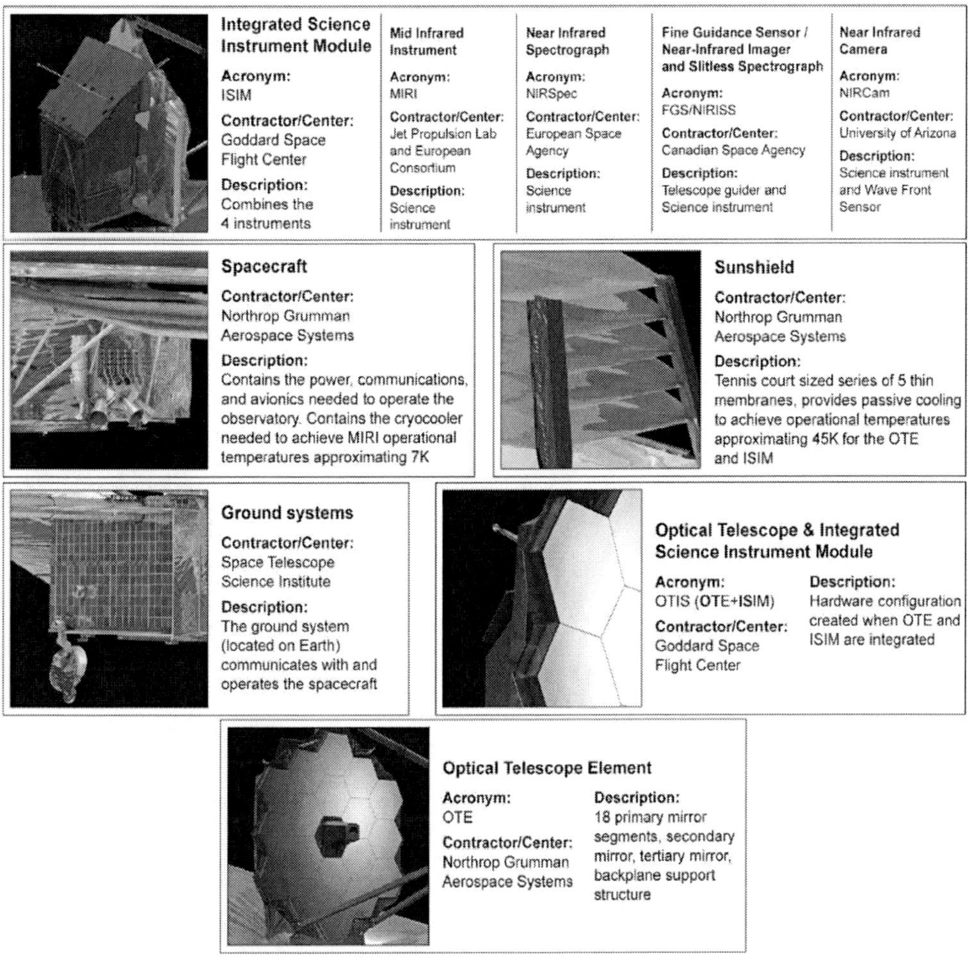

Source: GAO (analysis); NASA (data and images).

Note: The hardware configuration created when the Optical Telescope Element and the Integrated Science Instrument Module are integrated, referred to as OTIS, is not considered a subsystem by NASA, but we categorize it as such for ease of discussion.

Figure 8. Subsystems of the JWST: Interactive Information.

Appendix III: Our Evaluation of JWST's Cost Estimate Process

In determining that the National Aeronautics and Space Administration's (NASA) processes for developing the James Webb Space Telescope (JWST) cost estimate do not fully comply with best practices, we evaluated the project's cost estimation methods against our 2009 Cost Estimating and Assessment Guide. (See table 4.)

We applied the following scale across the four categories of best practices:

- Not met: NASA provided no evidence that satisfies any portion of the criterion.
- Minimally met: NASA provided evidence that satisfies less than one-half of the criterion.
- Partially met: NASA provided evidence that satisfies about one-half of the criterion.
- Substantially met: NASA provided evidence that satisfies more than one-half of the criterion.
- Met: NASA provided complete evidence that satisfies the entire criterion.

Table 4. Summary Assessment of JWST Cost Estimate Compared to Best Practices

Characteristic	Overall assessment	Best practice	Individual assessment
Comprehensive	Substantially met	The cost estimate includes all life-cycle costs.	Met
		The cost estimate completely defines the program, reflects the current schedule, and is technically reasonable.	Substantially met
		The cost estimate work breakdown structure is product oriented, traceable to the statement of work/objective, and at an appropriate level of detail to ensure that cost elements are neither omitted nor double counted.	Substantially met
		The estimate documents all cost-influencing ground rules and assumptions.	Substantially met
Well documented	Partially met	The documentation should capture the source data used, the reliability of the data, and how the data were normalized.	Partially met
		The documentation describes in sufficient detail the calculations performed and the estimating methodology used to derive each element's cost.	Minimally met
		The documentation describes step-by-step how the estimate was developed so that a cost analyst unfamiliar with the program could understand what was done and replicate it.	Minimally met
		The documentation discusses the technical baseline description, and the data in the baseline are consistent with the estimate.	Substantially met
		The documentation provides evidence that the cost estimate was reviewed and accepted by management.	Substantially met
Accurate	Partially met	The cost estimate results are unbiased, not overly conservative or optimistic and based on an assessment of most likely costs.	Partially met
		The estimate has been adjusted properly for inflation.	Partially met
		The estimate contains few, if any, minor mistakes.	Partially met
		The cost estimate is regularly updated to reflect significant changes in the program so that it is always reflecting current status.	Partially met
		Variances between planned and actual costs are documented, explained, and reviewed.	Minimally met
		The estimate is based on a historical record of cost estimating and actual experiences from other comparable programs.	Partially met

Characteristic	Overall assessment	Best practice	Individual assessment
Credible	Partially met	The cost estimate includes a sensitivity analysis that identifies a range of possible costs based on varying major assumptions, parameters, and data inputs.	Minimally met
		A risk and uncertainty analysis was conducted that quantified the imperfectly understood risks and identified the effects of changing key cost driver assumptions and factors.	Partially met
		Major cost elements were cross-checked to see whether results were similar.	Minimally met
		An independent cost estimate was conducted by a group outside the acquiring organization to determine whether other estimating methods produce similar results.	Partially met

Source: GAO analysis of NASA data.

End Notes

[1] GAO, GAO Cost Estimating and Assessment Guide: Best Practices for Developing and Managing Capital Program Costs, GAO-09-3SP (Washington, D.C.: Mar. 2, 2009); and GAO Schedule Assessment Guide: Best Practices for Project Schedules, GAO-12-120G (Washington, D. C.: May 30, 2012).

[2] The electromagnetic spectrum is the wavelengths of all the visible and invisible light. The infrared part of the spectrum, also known as radiant heat, has wavelengths that go from about 0.75 microns to a few hundred microns. The Hubble is designed to operate primarily in the ultraviolet and visible wavelengths of the spectrum from 0.1 to 0.8 microns. Humans cannot see in the ultraviolet region.

[3] James Webb Space Telescope (JWST) Independent Comprehensive Review Panel (ICRP): Final Report (Oct. 29, 2010).

[4] The hardware configuration created when the Optical Telescope Element and the Integrated Science Instrument Module are integrated, referred to as OTIS, is not considered a subsystem by NASA, but we categorize it as such for ease of discussion.

[5] A project typically reports to a program within a mission directorate, such as the Science Mission Directorate. A similar reporting structure was also instituted in the past with both the Hubble Space Telescope and the Mars Exploration Program when they began experiencing significant cost and schedule issues. For JWST, the change was made in response to recommendations from the ICRP.

[6] The JCL is a quantitative probability analysis that requires the project to combine its cost, schedule, and risks into a complete quantitative picture to help assess whether the project will be successfully completed within cost and on schedule. NASA introduced the analysis in 2009, and it is among the agency's initiatives to reduce acquisition management risk. The move to probabilistic estimating marks a major departure from NASA's prior practice of establishing a point estimate and adding a percentage on top of that point estimate to provide for contingencies. NASA's procedural requirements state that Mission Directorates should plan and budget programs and projects based on a 70 percent JCL, or at a different level as approved by the Decision Authority—which in the case of the JWST project would be the Agency Program Management Council—and any JCL approved at less than 70 percent must be justified and documented. NASA Procedural Requirements (NPR) 7120.5E, NASA's Space Flight Program and Project Management Requirements, paragraph 2.4.4 (Aug. 14, 2012).

[7] GAO-09-3SP.

[8] A work breakdown structure reflects the requirements and what must be accomplished to develop a program, and it provides a basis for identifying resources and tasks for developing a program cost estimate. The work breakdown structure should be used to define all program activities and tasks to ensure that the schedule encompasses the entire work.

[9] Earned value management is a project management tool that integrates the technical scope of work with schedule and cost elements and compares the value of work accomplished in a given period with the value of the work expected in that period. When used properly, earned value management can provide objective assessments of project progress, produce early warning signs of impending schedule delays and cost overruns, and provide unbiased estimates of anticipated costs at completion. As a best practice the work breakdown structure should

match the schedule, cost estimate, and earned value management system at a high level so that it clearly reflects the work to be done.

[10] A summary schedule is derived from the integrated master schedule and is often used in a JCL analysis when, for example, time or resources needed to simulate the full detailed schedule are limited.

[11] Level of effort activities are those activities that have no measurable output and cannot be associated with a physical product or defined deliverable. These activities are typically related to management and other oversight that continues until the detailed activities they support have been completed.

[12] An independent cost estimate is conducted by a group outside the acquiring organization that uses the same detailed technical information as the program estimate to determine whether other estimating methods produce similar results.

[13] GAO-12-120G.

[14] An integrated master schedule constitutes a program schedule that includes the entire required scope of effort, including the effort necessary from all government, contractor, and other key parties for a program's successful execution from start to finish. As a key focal point for management, the integrated master schedule documents and integrates the planned work, the resources necessary to accomplish that work, and the associated budget. It also consists of logically related activities whose forecasted dates are automatically recalculated when activities change allowing for management to respond to the consequences of these changes.

[15] NASA, Office of Inspector General, NASA's Management of the Mars Science Laboratory Project. IG-11-019. (Washington, D.C.: June 2011).

[16] Mass is a measurement of how much matter is in an object. It is related to an object's weight and is mathematically equal to mass multiplied by acceleration due to gravity. The project uses mass for JWST because when it goes into space, its weight changes with gravity, but its mass stays the same.

[17] NIRCam, NIRSpec, and FGS/NIRISS use detectors fabricated by the same vendor; whereas MIRI uses different detectors that do not need to be replaced.

[18] NPR 7120.5E, table 2-5 and appendix A (Aug. 14, 2012).

[19] Executive Order 13589, Promoting Efficient Spending, The White House, November 9, 2011. NASA provided allocations to the Centers based on this order. In turn, Goddard conducted an internal travel prioritization process and center management determined the allocation for JWST and other Goddard projects.

End Notes for Appendix I

[1] GAO-09-3SP. For the cost guide, GAO cost experts assessed measures consistently applied by cost-estimating organizations throughout the federal government and industry and considered best-practices for the development of reliable cost-estimates.

[2] NASA, 2008 NASA Cost Estimating Handbook, (2008).

[3] GAO-12-120G.

[4] NPR 7120.5E (Aug. 14, 2012) and NASA Procedural Requirements 7123.1A, NASA Systems Engineering Processes and Requirements with Change 1. (Nov.4, 2009).

INDEX

S

T

U

V